Power Quality in Renewable Systems

Divya S

CONTENTS

Page
No

ABSTRACT

New trends in power systems includes the placement of Distributed Generators (DGs) to overcome the draw backs of conventional power systems. Distributed Generator can be connected near to the load points. Hence the placement of Distributed Generators is an important factor to be considered for the analysis due to its positive as well as negative impacts.

More concentration has to be given on renewable energy sources and its penetration level. The new trends in power system includes the renewable energy sources with their large penetration levels interconnected to Distribution Systems. The penetration level of renewable distributed generators plays an important role in the interconnected system and their impacts (both positive and negative) can be considered for the analysis.

The Positive impacts includes improvement in voltage profile, reduction in active and reactive power losses, thereby improving the power quality and reliability of the system. The improper placement i.e.; penetration level, type of Distributed Generators and location of Distributed Generators leads to the increase in active power and reactive power losses, creates variation in voltage profile and also there of chances of reversal of power flow.

In this dissertation, renewable distributed generators (Solar and Wind) for different penetration levels are considered to improve the system performance which includes the quality of the power to be supplied. Standard IEEE 33 bus, 69 bus and Practical (Mysuru Zone) 32 bus test systems are considered and tested for heavy load condition. The integration and behavior of the Solar, Wind and hybrid Distributed Generators are analyzed with analytical, intelligent nature inspired technique and a meta-heuristic hybrid Technique for different penetration levels of the integrated Distributed Generators

A novel analytical method for integration of Solar, Wind and Hybrid Distributed Generators at different penetration levels are considered in Chapter 2. The developed algorithm includes different indices such as Logarithmic Value of Voltage Deviation Index (LVDI), Voltage Profile Index (VPI), Voltage Profile Improvement Index (VPII), and Weighted Sum of Voltage Profile Index (WSVPI). The penetration level of Distributed Generators is selected based on percentage penetration levels i.e.; 10%, 20%, 30%, 40% and so on.

In Chapter 3, a modified nature inspired Bio Shuffled Frog Leap Algorithm is used to obtain the optimal location of Solar, Wind and Hybrid Distributed Generators at different penetration levels based on different power quality assessment indices such as number of buses experiencing voltage sag (Nsag), number of buses experiencing voltage swell (Nswell), system average RMS frequency index (SARFI) and system average RMS frequency improvement index (SARFII). The main objective function of this Chapter includes minimization of active power losses, number of voltage sag and voltage swell buses, and minimization of Voltage Sag Index (VSI). The analysis is carried out using Logarithmic Value of Voltage Deviation Index (LVDI), Power Loss Reduction Index (PLRI), System Average RMS Frequency Index (SARFI) before and after placement (integration) of DG and System Average RMS Frequency Improvement Index (SARFII). Also, Constant Power Load Model is considered for the case of heavy load condition. Drastic changes in the test system can be observed when the system is over loaded. This problem is solved by using Modified Shuffled Frog Leap Algorithm (MSFLA). The enhancement of power quality is achieved by nullifying the effect of Voltage Sag and Swell at the affected buses by using the indices based on the variation of voltage.

A Hybrid approach combining the advantages of both Modified Shuffled Frog Leap Algorithm (MSFLA) and Particle Swarm Optimization Algorithm (PSO) (MSFLA-PSO Algorithm) is proposed in Chapter 4 for the optimal location of Solar, Wind and Hybrid Distributed Generators at different penetration levels. Also, the optimal placement (Size and Location) of Solar, Wind and Hybrid Distributed Generators is obtained and compared with the fixed penetration levels. The obtained results using this algorithm is compared with the results obtained in Chapter 3 and also with other different Algorithms selected from the Literature Survey. In this Chapter, optimal location for different penetration levels (10%, 20%, 30%, 40%, and 50%) is obtained using the Hybrid Algorithm.

Further analysis is carried out for optimal placement of Wind Distributed Generator by considering fixed active and reactive power (Fixed P, Q), fixed active power-optimal reactive power (Fixed P-Optimal Q), fixed reactive power-optimal active power (Fixed Q- Optimal P) and optimal active power-optimal reactive Power (Optimal P- Optimal Q). Standard IEEE 33 bus, 69 bus and Practical (Mysuru Zone) 32 bus test systems are considered to show the accuracy of the proposed algorithm

LIST OF FIGURES

CHAPTER 1

1 INTRODUCTION

1.1 General

The electric power industry is the largest consumer markets in the world. Due to the deteriorating conventional sources, the industry is in need of alternatives. There is a requirement to supply the power at appropriate voltage, whereas research shows that as much as 13 percent of the power which is generated is always mislaid at the level of dispersal and the ratio X/R is always low for the distribution level when contrasted with transmission [5]. These days, when referring to a defined quality product, since electricity is intangible and transient in nature electricity is rather considered a unique product.

The DGs which stands for the Integration of Distributed Generations to various power systems also bring many benefits and some of the key benefits include reduction in transmission and resources of distribution system, reliability increase, better stability and the superior power quality. All these benefits however depend on the configuration of a system, its capacity for DG and its type. Some other factors involved are integration location and DG management. For DG the optimal allocation is always needed for the power system with additional benefits. Power quality is a major issue to be considered for the analysis in power system when distributed generators are integrated.

Operation of Renewable distributed generators (RDG) will increase the rate of power quality issues, hence extra care has to be considered when RGD's are integrated to the grid . Analysis is carried out on variation of voltage profiles, varying load, stochastic wind power and power losses.

1.2 Current power scenario

The power sector of India is among the most diversified sectors of the world with a wide range of conventional and non-conventional sources. A massive increase in generation is required to meet the rising demand for electricity. For the conventional sources the target of electricity generation from the last two years, 2019-2020 has been allocated to be as 1330 Billion Unit or BU. This indicates that as compared the previous years 2018 and 2019 there is over 6.46 percent more conventional generation of power.

India stands fifth for power generation capacity and third in terms of production in the world with 1330 Billion Units (BU) during the fiscal year 2019.

The renewable energy sector added a capacity of 8,532 MW to the grid, just about the 55 per cent of its target of 15,602 MW for 2018-19. In the previous two years, the renewable sector added new capacity in the range of 11,000-12,000 MW per year, supported by a boom in the solar power sector. The power generation from 2014-15 to 15-16 is slightly reduced and later on in the coming years it keeps decreasing due to the fact that the difference in power generation is very small compared to the previous years [fig 1.1(b)]

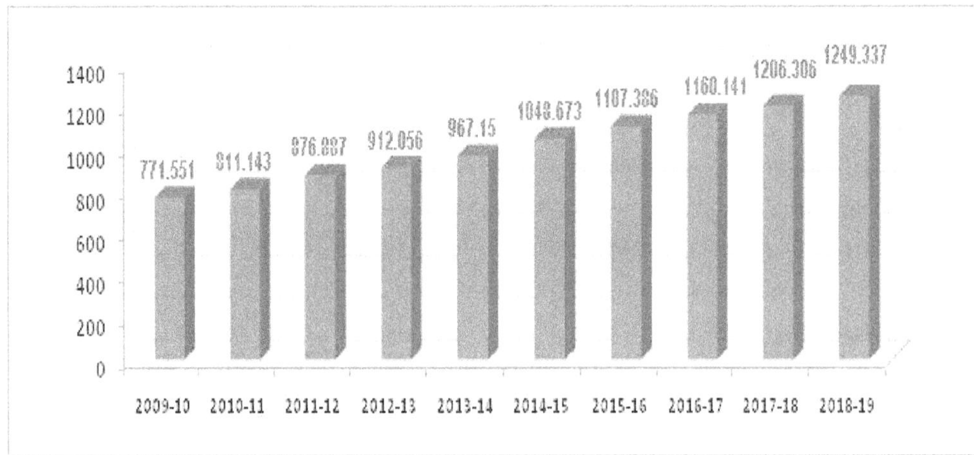

Figure 1.1(a): Power Generation (Billion Units) in India from past ten years

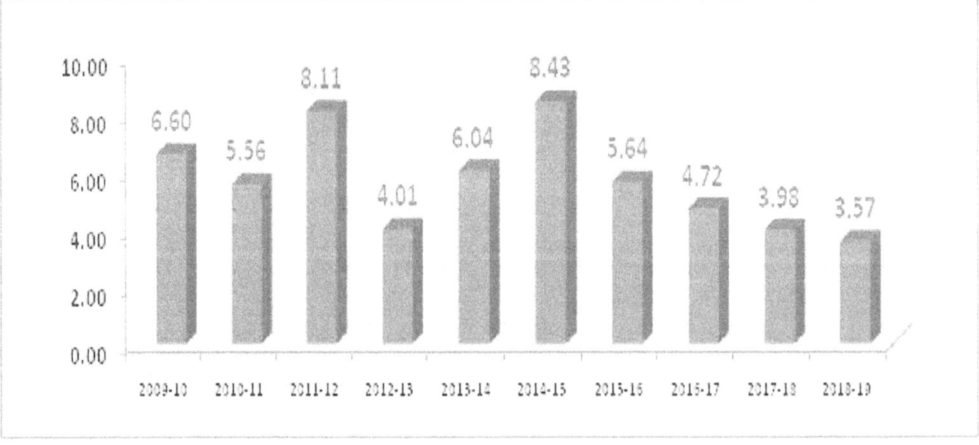

Figure 1.1 (b): Power Generation growth in India from past ten years (in terms of percentage)

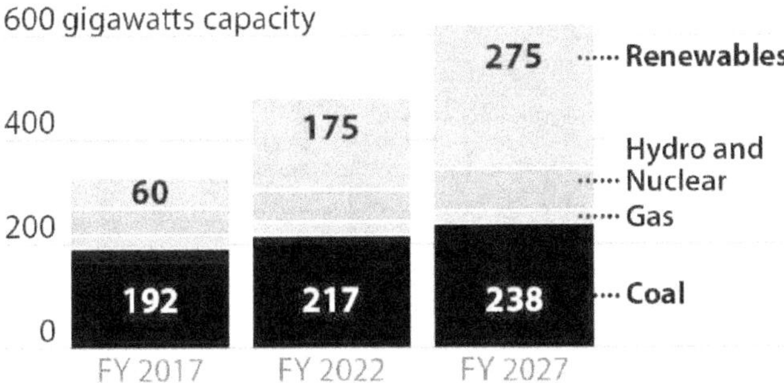

Figure 1.1 (c) : Growth in Renewable Energy according to National Electricity Plan

1.3 Distributed generation

The installation and execution of reduced and modular technologies for the generation of power is known as Distributed generation, it is abbreviated as DG. It can be combined with storage as well as energy management systems. The operation of electricity delivery systems at or near the end users is improved with the help of this DG. Electrical grids are sometimes installed with these systems as well while in some cases DGs are not connected with these electrical grids.

A range of technological options can be employed by the distributed generation from various energy resources such as renewable and non-renewable and in connected grid or off grid mode they are functional. Distributed generation system size is also variable. The size ranges from less than a kilowatt to few megawatts.

1.3.1 Technological options

On the basis of prime movers which are used for example engines, fueling cells, turbines the DG operations can be classified. They are also classified on the basis of renewable or non-renewable fuel resources. In the countries like India, in the distributed generation projects many renewable energy technologies are being employed. These technologies are of various kinds such as photovoltaic systems, small wind mills to run wind turbines and this is also known as aero generators, biomass gasifiers, smaller hydro power plants etc.

The figure below mentions the options of technology available for the distributed power generation.

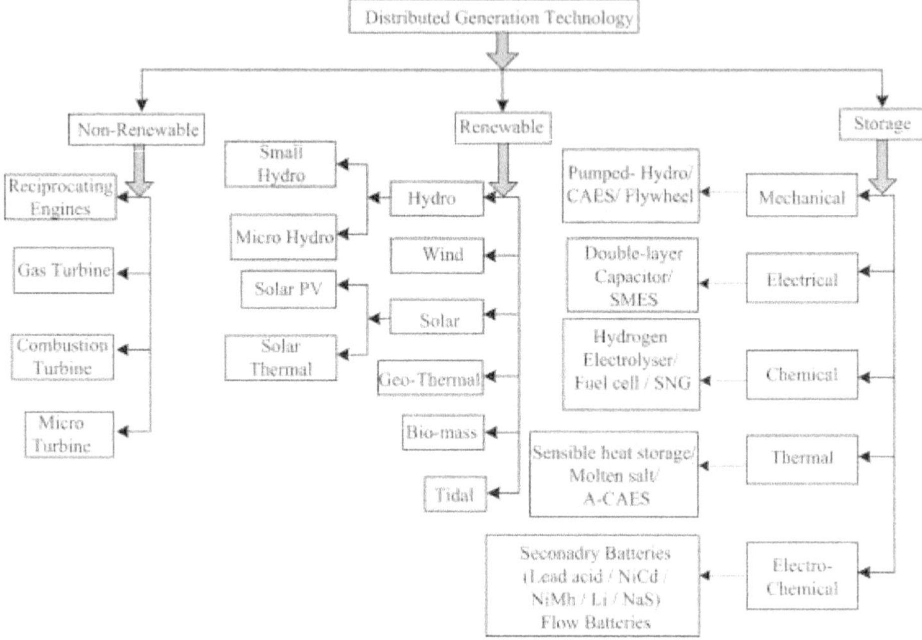

Figure 1.2: DG Technologies

1.3.2 Distributed generation and its relevance in India

In some countries like India, three distinct markets are found in the distributed generation.

- Small power generation systems back up which also include the generator diesels being utilized in the small commercial as well as domestic units.
- For electrification of rural and remote areas the stand alone off grid systems and mini girds.
- Some larger and captive kind of power plants is installed by the industries which are power intensive in nature.

  There are some issues which are needed to be addressed by the distributed power generation however and these are mentioned below:

- Load shortages of high peak: the distributed generation systems with the deficit of 12.3 percent is seen the highly effective solution to the problem regarding reduction of peak demand.

- Higher losses in transmission and distribution: 20.33 percent of the total energy which is available is the current loss. These loses can be tremendously reduced by the distributed power generation systems and hence the gird network and its reliability increases.

- The areas which are not easily accessible and are remote: since, India is a big country and all the areas are not easily accessible, it is not feasible to extend the grids as it will cost a lot. A major role can be played by the distributed generation in this case.

- Electrification of rural India: the electrification of rural India is considered as a foremost priority of Indian Government and in this case, where less feasibility is there for the gird extension the facilities such as decentralized distribution generation are being directed by Indian government along with local networks for distribution.

- To new and modern power demands the fast response: the distributed generators have modular nature and hence the system is often coupled with periods of lower gestation and this allows the additions of easier capacity whenever needed.

- Power quality and system reliability enhancement: there are some disruptions as well which include the failure of grid, and these can be prevented by producing electricity closer to the consumer. The maintenance of power quality, frequency and voltage is easier as well.

- Load management and better energy possibility: the possibility of combining energy store and management systems is offered by the distributed generation systems.

- Existing grid assets and their optimal utilization. In the distribution network the inadequacies are considered as the foremost reason for the low quality or poor supply of power.

- The optimal use of grid is facilitated by the distributed generation and in this way the network congestion is reduced and the reliability of the grid network increases.

1.3.3 The distributed generation in context to the policy

The government of India has its own integrated energy policy for the planning commission. It envisions the security of energy for the states and locals by mentioning that the energy produced is supposed to be safer and more reliable, viable in terms of technology and economy and sustainable as well keeping in mind its various forms and energy fuels which are conventional and some alternate sources of power production.

In 2003, In India the Electricity Act was introduced and it has given thrust to the system of distributed generation especially in terms of electrification of rural areas. According to this act, along with the extension of grids as a model for the electrification of rural areas of India, the conventional and renewable energy systems are used by the distributed system. As alternate modes of electrification of rural electrification, the distribution of electricity is also done via NGOs, units of local government and community groups and franchisees of distributed utility.

Along with this, as per this act the new projects launched by the person or extension of currently existing infrastructure for the generation schemes and distribution are given exemption from any license related obligation.

On 12th February 2015, the National Electricity Policy was notified and under the section of 5.1.2 (a) under the rural electrification component which offer a system of more accurate and reliable rural electrification, by the extension of lines of transmission the rural electrification distribution backbone be established. On the other hand, where feasibility is lacking as per 5.1.2 (d), decentralization of the facilities regarding distribution generation along with local distribution networks are supposed to be transmitted making use of conventional and non-conventional energy sources.

According to the Electricity Act 2003, the compliance of sections 4 and 5, the rural electrification policy is supposed to be designed by the central government. In section 3 (3.3) the policy also recommends the electricity decentralized and distributed generation and this is done by the networks of distribution at local level facilities set up so that other methods of generation are used which are either conventional or non-conventional for the generation of power. The government of India launched two different schemes for this purpose, and these are RGGVY which is an abbreviation of (Rajiv Gandhi Grameen Vidyutikaran Yojna) and RVE which stands for the remote village electrification scheme. Both these schemes offer more than 90 percent of capital subsidy making use of DG (decentralized distribution generation) for the rural electrification projects. These all options are based on the fuels which a are either conventional or non-conventional.

1.4 Renewable distributed energy sources

Since we know that electricity in intangible and transient in nature, hence now days it is defined as a unique product with well-defined quality. There have been many concerns regarding the impacts on the power supply quality with the electricity markets liberalization and the steady increase in the penetration of the Distributed Generation from RES (renewable energy sources). These are driven by the policies which are environmental friendly and have some regulatory frameworks as well. To ensure high level of power supply quality close attention is supposed to be given to ensure that distributed generation or DG is integrated securely and successfully into the systems of electrical power.

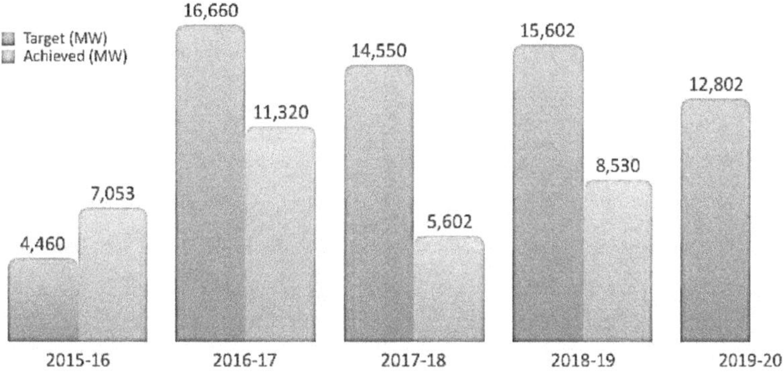

Figure 1.3 (a) : Renewable Energy Targets and Achievements of India

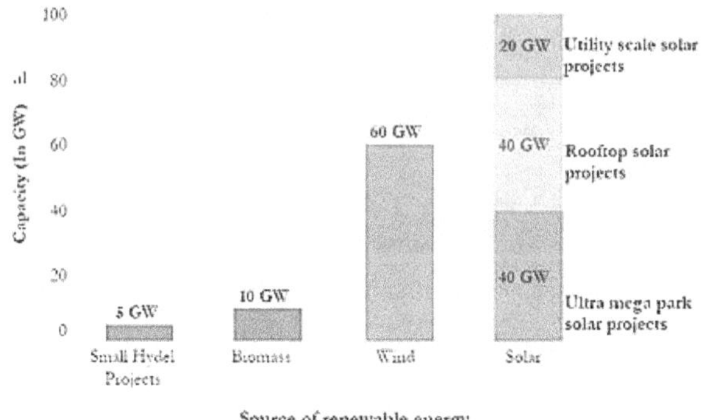

Figure 1.3 (b) : India's 2020 Renewable Energy Target

1.4.1 Renewable energy and its higher penetration

The distributed Generations (abbreviated as DG) with higher penetration are now days new challenge for the systems of electrical power used traditionally. From DGs power injections change the flows in the network power to modify the loses in energy. DGs are considered to be useful in the reduction of loses but this is not always the case. Keeping in mind the Distribution generation technology, it is important to note that in terms of loss reduction wind power is the one which shows the worst behavior. With reactive power control the distributed generation units offer better profile of network voltage and minimize the losses.

1.4.2 Impacts of higher penetration distribution

The distributed generation systems are traditionally designed in such a way that they work in the radial fashion from the substation source to the load with flow in one direction. The Distribution generation or DG has been appearing more often on the system of distribution.

In the recent past, due to the enhancement of viability of economy, commissions for the public utility and incentives needing the DGs consideration as substitute for the upgrading of the traditional circuits and portfolio standards for the state renewable sources, the DG poses the design and challenges in execution include the higher level of penetration which is unpredictable mostly and the output sometimes is highly variable as well which indicates a challenge that is lesser familiar.

1. Impacts related overloads: the capacity ratings of the circuit elements tend to get exceeded in number of ways due to higher penetration of renewable systems. Most intuitively perhaps, from the attached systems the net generation is likely to overload the elements of the circuit situated between the systems of generation and load centers in the circuits under discussion.

2. The impacts regarding Voltage: PV due to higher penetration along with wind are likely to have their impact directly in number of ways at the circuit voltage. In solar PV the rise and variations in voltage are caused due to fluctuations and these are regarded as two most potentially problematic impacts of PV and wind. When large amount of solar PV and Wind are linked near the lightly loaded feeders end these effects are pronounced specifically.

   From the system of PV, the real and reactive power can impact the current voltage, output rise and fall, voltage fluctuations in the steady stare on the given circuit. In this way there is a strong impact of all this on the quality of power and device operations to control voltage.

3. Impacts of Reverse power flow: On the distribution system upstream the power flow in reverse for the wind system and PV is likely to occur during higher generation and lighter load. For the protection system as noted previously the reverse flow can cause some problems and also for the regulators of voltage.

   One of the significant impacts due to the DG is the reverse power flow (RPF), which generally occurs when the generation of a distributed electric power plant exceeds the local load demand, causing power to flow in the opposite direction to normal. This phenomenon can be produced by the intermittency in the renewable energies, which depends on climatic factors. The consequences can be evidenced in the power system with voltage peaks; therefore, the sensitivity and various parameters of protection coordination are critically affected, which has further implications in the power quality.

4. Impacts of system protection: for the renewable DGs higher penetration is likely to change the levels of fault current and in this way, review is essential for the distribution network's protection and to manage its currently execution. High penetration has a lot of impacts for the distribution system protection for the renewable DGs and hence for the analysis these all impacts are supposed to be considered. The solar PV (photovoltaic) connected to the grid and the DG powered by wind often brings a lot of advantages and challenges to the distributed system. In the solar distributed apps, electricity is generated by small PV systems (5-25kilowatts kWs) and this is done for the consumption at site and with lower voltage transformers it is connected on the systems of electricity utility. The transmission line losses can be reduced by deploying the distributed PV, it also increases the resilience of gird, minimizes the costs for generation, also it reduces the needs of investment in modern capacities for the utility generation. When the suitable tools are used and there is proper calibration, the issues of reliability are mitigated by the distributed systems by offering standby capacity in the period of outage periods and the time of utility disturbances.

   DGs and their higher penetration on the system of distribution can end up in the impacts regarding reliability which are directly and indirectly linked with the voltage, power overload, reverse flow of power, configuration of a circuit etc. Potential impacts can be analyzed and also there is a need of following mitigation techniques for the integration of DG for the larger penetration of the current power system.

1.5 Power quality

 Power quality issues are of a great concern since the advancement in the electronic technology leading to the change in nature of electric loads, due to their non-linearity they lead to disturbance in the voltage waveform. The most sensitive areas for the requirement of good quality of power are continuous process industries and information technology service sectors. Any disturbance occurring will results in huge financial loss and consequent loss of productivity. Hence, some steps must be taken to mitigate these issues to obtain good quality of power.

1.5.1 Types of power quality problems

1. Sag in voltage: When the system voltage reduces to 10 to 90% of the RMS value of actual voltage for 0.5 second to 1 minute is called as the sag. This is caused mainly due to faults or overloading due to large motors. The voltage sag results in malfunctioning of protective devices which leads to an interruption in the processes and tripping. The examples of voltage sag are shown in Fig.1.4 (a) and fig 1.4 (b)

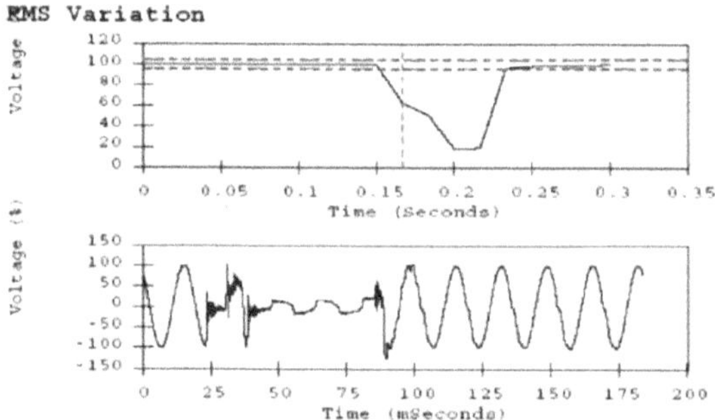

Fig.1.4 (a) : Instantaneous voltage sag due to fault

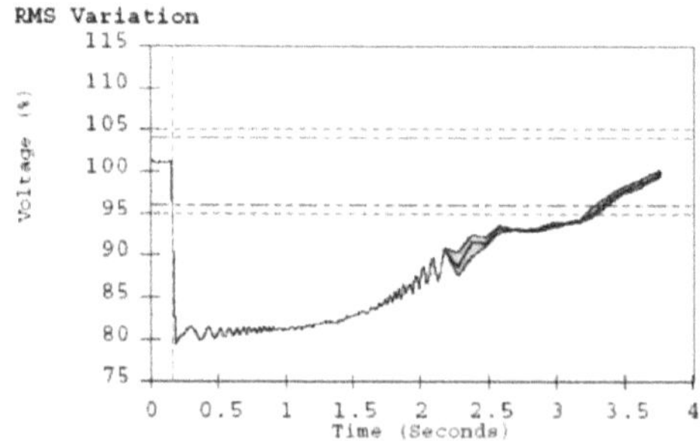

Fig.1.4 (b): Voltage sag due to starting of the motor

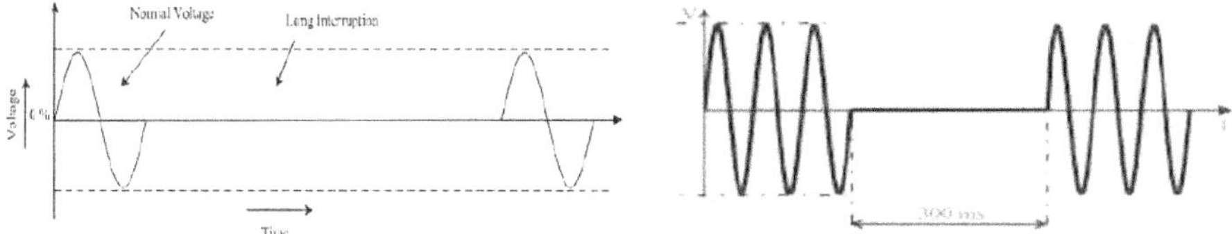

Fig.1.5 (a) : Long and short interruptions in supply voltage respectively

2. Very short interruptions: The interruption of supply for very small time is called very short interruptions which are mainly due to the switching of protective devices resulting in tripping and malfunctioning of the equipment.

3. Long interruptions: The interruption of power for a long time which is higher than 1 to 2 sec is known as long interruption which is caused due to failure in the network or bad coordination of protective devices.

4. Voltage Spike: The fluctuation in voltage for the time between micro and milliseconds is known as a spike in voltage resulting in failure of components. The spike may reach thousands of volts in such cases. The main causes for these are lightning, switching of lines or capacitor banks, etc.,

5. Voltage swell These interruptions are defined as a sudden increase in voltage at a nominal value of frequency out of tolerable limit. The consequences of this are data loss, flickering; stoppage of sensitive equipment, etc., an example of a voltage swell is presented in Fig.1.5 (b).

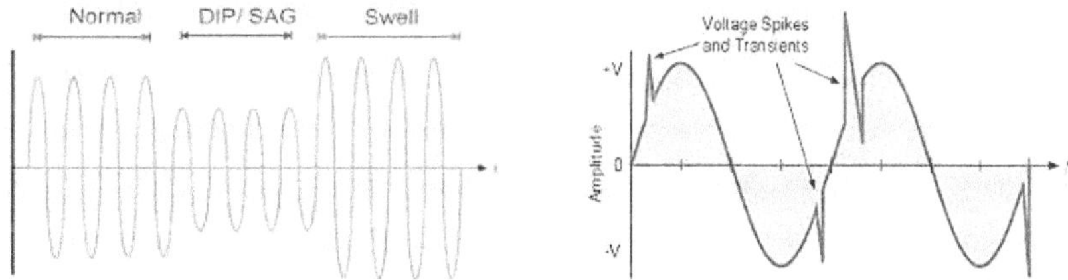

Fig.1.5 (b): Voltage swell and Voltage spike respectively

6. Harmonic distortion: The sum of sine waves with different frequencies which are multiples of original frequencies are referred to harmonics. The causes include the

welding machines, dc brush motors, arc furnaces, which leads to increased chances of occurrence of resonance, natural overload in 3-phase systems and interference in communication systems. Example of harmonic distortion is shown in Fig.1.8.

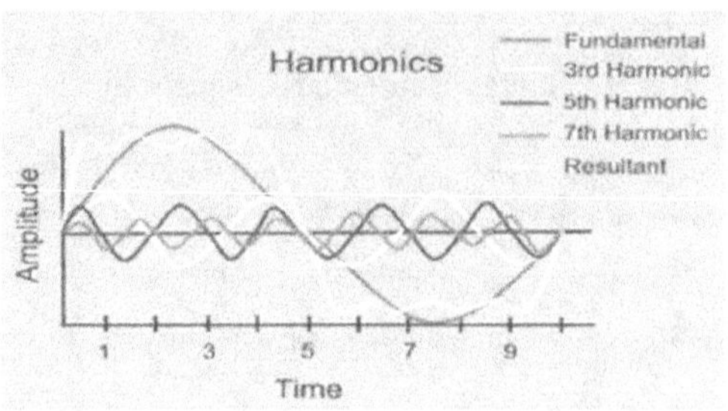

Fig.1.5 (c): Harmonics in the waveform

7. Voltage fluctuations: Amplitude modulation of the voltage with 0 to 30 Hz frequency signal can be said as the voltage fluctuations. The causes of which are the arc furnaces, oscillating loads and frequent start/stop of electric motors. These fluctuations result in under voltages, flickering of lights and screens and giving an impression of unsteady visual perception.

1.6 DGs and their optimal placement

There are two different perspectives to view the optimum allocation of distributed generation. From the point of view of independent producer, the benefit is to be optimized and this is the ultimate goal. Some constraints will be imposed by the utility to the units of generation when they are connected or linked directly to the system with high rated power, by default it can be assumed that to the system at any node the units can be connected. Hence, the approach for optimization in general would be the study of feasibility and to check the installation viability and select the size of economy regardless of any location. The control strategy would be used in case of dispatch able units. It will provide maximum benefits as well. The goal is to maximize the distribution generation positive impact from the utility point of view. For example, the supply of voltage, losses of energy as well as deferring investment while it also reduces or tends to avoid the negative impacts on the performance of a system, these include protective tools mismanagement and lack of coordination, during lower periods the risks of over voltages etc.

1.7 Literature review

On the electrical network there are a lot of beneficial impacts of distributed generators DGs although a lot of concerns are involved with them. On the allocation and effect of DGs on the system there has been a great amount of research and the aim is to offer a better and superior voltage profile and reduction in the actual loss of power. [1].

The optimal allocation is a study area giving numerous techniques for placement of DG and size of DG's [1-34, 45-50]. The more focused part of the research was reducing losses. Also, the factors like "voltage profile, reliability, DG capacity, and power management, so on". These are considered for minimization or maximization of the considered single or multi-objective problems under different cases. The DG's must be reliable for operation, hence they have to be strategically placed to give the enforcement in the grid and improve the system reliability [35-43]. The size of DG is less than 25% of the total system capacity could result in the overall system stability. It is seen that all the approaches are addressed with a particular loading condition but in real time bases the loads are dynamic in nature, hence may produce inappropriate results when analyzed using these algorithms.

Many techniques of optimization such as evolutionary algorithm, ant colony technique, and particle swarm optimization are given in [54-71]. In [57] PSO has been used for optimal allocation of DG's and the obtained values were compared to that of the analytical approach. The techniques such as prime dual interior point method [58], mixed integer nonlinear programming [59-61], EP technique [62], analytical technique [63-67], LP technique [70], genetic algorithm [71], meta-heuristic approach [72], etc., have been employed to solve the DG allocation problems in the network.

DG's are connected to the electrical network at their load points on the customer site of the meter [63]. It acts as a backup source during the failure of the main supply, also during the unavailability of the power as well as an additional source during the normal operation to reduce the system losses [2]. Integration of DG to the system under operation is an art and important from the point of considering system losses, power quality, the voltage variation at different points [70]. It is a practice to use analytical method for optimal placement of DG in the initial cases, the optimal placement of DG on the power grid to reduce the total system losses by improving voltage profiles to increase the capacity of the system was proposed [71].

Optimizations techniques are utilized to maximize the advantages posed by DG's as they give appropriate voltage profiles and lower the losses in the network [8-10]. The study has shown many heuristic techniques proposed to solve the optimization problems. The paper [10] shows that the constant load model determines the effect of different loads and the sensitivity to frequency and voltage. In the paper [11], the authors have proposed an analytical method to optimally allocate the DG's which used the BIBC (branch injection to branch current) and BCBV (branch current to branch voltage) matrices to get the load flow reducing the search space. The authors of the paper [12] present the application of dynamic programming for placement of DG taking into account the voltage improvement and loss reduction.

The study of the heuristic method was followed by the evolutionary techniques [13-15] to find the optimal location of the DG. The authors in [15] proposed an adaptive particle swarm optimization (APSO) to place multiple DG's where the objective was to minimize the real power losses only. The most sensitive busses are identified for the placement of the DG in [16], same capacity DG is installed at all the buses and analytically calculates the lowest loss combination as the best size-site pair. A GA based algorithm is proposed in the paper [13] to install multiple DG's for cost minimization.

Optimal Integration of Dispersed Generation to improve the system performance using different techniques from the beginning to the Recent literature

To solve the optimization-based problems regarding DG the mathematical algorithms are used. there are three main groups under which these algorithms are divided:

- The Conventional based methodologies like linear and nonlinear programming, Alternative current's optimal flow of power, continuous flow of power, MINLP which is an abbreviation of mixed integer nonlinear programming and the technical and analytical approach.
- Simulated Annealing is included in intelligent searches along with various evolutionary algorithms (EAs), TS which is an abbreviation of Tabu search, PSO which stands for particle swarm optimization, ACSA which is an abbreviation of Ant Colony System Algorithm.
- To address the technical and economic risk the fuzzy set their which is mentioned as FST is used.

1.7.1 Conservative methods

1. LP which stands for Linear programming is utilized in order to solve models of ODGP in [21] and [22] for the attaining of maximized DG penetration and maximised harvesting of DG energy respectively.

2. NLP is an abbreviation of non-linear programming; it is mixed with MINLP which stands for mixed integer non-linear programming. It is a model for probabilistic generation load with all the execution conditions possible and is reduced into the deterministic model which is used to get a proper solution in the MINLP which stands for the mixed integer no linear programming technique for the allotment of either DG operational units [23], or different DG units -24]. As a multi period alternative current control the ODGP is formulated and this is done to solve the NLP [25][26][27]. For the connection of DG the distribution network capacity by OPF formulation is computed which with the help of interior point method is solved [28]. For the optimal allocation of many different kinds of DG units MINLP is applied and in doing this the fluctuations of electricity market prices are considered [29]. In the hybrid electricity market, an ODGP is applied and this is done for the estimation of MINLP [30]. The solutions regarding optimization are made making use of models of integrated distribution and planning and the ODGP implementation as an alternative option [31].

3. Analytical methodology: with the load distributed uniformly at a radial feeder, this is a special analytical method and is also known as 2/3 rule. The installation of 2/3 DG is also suggested in it for the 2.3 length of time in the incoming generation capacity. On the other hand, this special method remains lesser or non-effective for the loads distributed uniformly. For the optimal location of a single DG is introduced in a fixed size with these two analytical methods. The method number one is applicable to the radial power systems while the second method is considered appropriate for the meshed power system. In the analytical methodology which is used, the formula of exact loss is suggested to optimize the site and size of DG [34]. With the help of loss of sensitivity factor the analytical method is used on the basis of injection of an equivalent current and this is done to determine the optimal size and single DG location. In [36] the proposal is given for the analytical approach to determine the multiple DGs optimal locations in collaboration with the special algorithm known as Kalman filter algorithm and this is used to determine the optimal size. In [37] for the optimal location and size of one or two DGs are suggested for the analytical expressions. For the determination

of optimal size as well as DGs power factor of various types the analytical expressions are proposed in the section [38].

4. CPF which stands for Continuation Power Flow: it is another method which is used for the DG based placement on the power flow analysis and to determine the voltage collapse most sensitive buses and in 2008 it was suggested by Heydayati. As per the mentioned procedure, the bus to voltage collapse which is most sensitive and the maximum load is found by the working of the flow program for continuous power. Once the sensitive bus is determined, one unit for DG is installed on that particular bus with the certain capacity. Once the DG unit is installed, there is an execution of program for power flow and calculation is made for the objective function. When there is an inappropriate estimation of an objective function, till the objective function is estimated the algorithm would iterate.

1.7.2　Methods based on Intelligent Searches

The intelligent searches are based on heuristic methods and in DGP these are implemented so as to deal with the problems regarding local minimum and various uncertainties. With fuzzy set theory and other methods of conventional optimization these methods are combined and this is done to solve the problems regarding DGP [39].

1. SA which is an abbreviation of simulated annealing: It is a process whereby the simulation is done for the process of annealing for the optimization of problems. Escaping local minima potential, it has is done by the probability function in new solutions acceptance or rejection. Kirkpatrick, Gelatt, and Vecchi in 1983 (Vidal, 1993) introduced this SA approach. And since the outcomes are good with this method, it gained much popularity in mid of 1980s and also the implementation of this model is simple. (Rao-Sepulveda et al, 2003).

One of the famous experts named, Dr T Ananthapadmanabha made use of this technique known as SA for the reactive power consumption for global optimal solutions. For the proficient system by affording the primary estimate and the reactive sources location the SA technique was made suitable making use of another method known as SVM. SVM is an abbreviation of set voltage magnitude method. For the modification of step size for the control variables the set of rules were established in the time when in control direction the switching was sensed. For the optimal reactive power control, the general rules were developed. For the practical 85 bus radial distribution system the development of proficient system was tested as well in the South Indian city, Mysore.

Authors presented a model in (utthibun et al, 2010), for the determination of DG size and optimal location for the minimization of loss in electricity, loss in emission and contingency making use of optimal tools such as SA. The initial temperature and procedure for cooling is important for the SA utilization. On the initialization, perturbation, schedule for cooling as well as probability acceptance the algorithm is generally based.

2. EAs are also known as evolutionary algorithms which are somewhat different from other methods of optimization used in conventional forms and in cost functions and constraints it does not need any differentiation. The evolutionary algorithms are also based on the solutions of global optimization and process of optimization by a finite number of evolution steps which are performed for the probable set of solutions. (Goldberg, 1989; Pham et al, 2000). There are some artificial intelligence methods used for the optimization based on natural selection which include mutation, genetic recombination, crossing over, reproduction, natural selection etc. these methods are evolutionary programming or EP, evolutionary strategies ES, and genetic algorithm GA. A candidate is perturbed randomly by mutation; the parts of them are mixed randomly for a novel solution by recombination; the choosing of random position is involved in crossing over in two different strings and after this position the strings are swapped a bit. In the same way replication of the successful solutions takes place in reproduction and solutions are determined in the population. The poor solutions however from the population are washed out via selection.

3. Many similarities are shared by this method. At first the EP is introduced, it is then followed by ES and GA (Goldberg, 1989; Lai et al, 1996). For EAs the simple and enhanced versions are there which have been successfully implemented for GDP in literature keeping in mind the single as well as multiple objective functions to somewhat variable constraints. To solve the optimal sitting and sizing of distributed generators probability is there ad this is done via GA as mentioned by (Silvestri et al, 1999). Enhanced Hereford Ranch Algorithm and GA are variants of GA which are suggested in [40] for the optimal sizing of DGs. For the solution of problems of OGDP the GA is applied along with some sort of reliability constraints in [41]. The variable power concentrated load models are used in GA for the optimal solutions of ODGP [42] along with various loads of distribution and concentrated loads for constant power [43] [44].

Another approach which is based on value is also considered since it has a lot of benefits and DGS costs are involved as well. This is developed and solved with the help of GA and the factors considered in this are optimal number, size and location of the DGs.[45]. The multiple objective ODGP is transformed into the OGDP single objective which is solved with the help of method known as GA. [46]. There is further applied a theory of decision for the solution of problems regarding ODGP under the power quality issues of uncertainty. [47]. In ODGP model a fuzzy GA is used and it also reduces the cost of power loss [48]. For ODGP the hybrid GA and fuzzy goal programming is used [49]. In [50] the tabular and combined GA is suggested. The weighted multiple objective ODGP models are solved in GA [51]. It has been emphasized by the authors that GA is combined with the flow of optimal power to offer the sites with best combination in the networking distributions for the connection of DGs in predefined numbers. In (Ochoa et al, 2008), based on dominating sort of genetic algorithm NSGA a multiple objective programming approach is used and it is applied to determine the configuration needed for the DWPG maximum iterations. The DWPG stands for distributed wind power generation and in this voltage as well as thermal limits are satisfied.

4. The TS which is an abbreviation of Tabu Search is an ODGP problem which with the help of TS methodology is resolved especially when there is a case of loads and their uniform distribution. [51]. ODGP and optimal placement of reactive power sources is solved by the TS at the same time. [50].

5. PSO stands for the particle swarm optimization which is applicable in order to resolve the models of ODGP system of distribution along with non unity powers keeping in mind the models of power loads. [5]. PSO in its enhanced and improved version is suggested for the DG types of optimal placement injecting the real power and absorbing the reactive power. [52].

6. The optimization of Ant content is suggested as well and this is done to solve the problems regarding ODGP [53].

7. ABC is an abbreviation of Artificial Bee Colony and in this special method there is a tuning of only two control parameters as suggested in [54].

8. DE is an abbreviation of Differential Evolution and the computer based optimal DGs and these are incremental bus voltage sensitivities and sizes of DGs are estimated with the help of DE [55].

9. The FST is a fuzzy set theory, is a concept which was initially launched as a routine device in order to deal with the soft modeling and uncertainty and in power systems it

is now days used extensively. [57]. the modeling of fuzzy variable is done by the membership function which assigns membership degree to adjust it well. Zero to one is the degree of variation of this membership.

10. The SLFA is known as a shuffled frog leaping algorithm which is actually meta heuristic to handle the problems of large-scale optimization. [105]. On SLFA in the previous researches there have been many efforts which denote the basic SLFA next generation along with diversification of SLFA modified or hybrid SLFA. Attempt is made for these structures to highlight; there are some benefits and enhancements of this structure.

1.7.3 MSFLA, an abbreviation of Modified Shuffled frog leaping algorithm

There are three main phases whereby the basic SLFA Meta heuristic techniques are structured. [100]. these three main phases are called initialization, evaluation, shuffling. In the phase of initializing there is a random selection of f frogs. In the phase which is known as evaluation the descending order of witness value is used to sort the frogs. The m groups are used further to sort the partitioned frogs which are able to carry out the independent local search. There is an improvement of all the memeplex so that in the end the better value of fitness is obtained. To raise the value of fitness is feasible due to frogs of high quality and for the low-quality frogs it is easier to reduce the fitness value on the basis of several predefined objectives. The phase of shuffling is there as well and global search is enhanced via it. [101]. After the accomplishment of evolution in this phase, there is a shuffling of memeplexes. There is a global optimization of all the frogs and this is done to generate new memeplexes. [102]. Until the best solution is formed the evolution as well as shuffling processes continues for many times.

For the allocation of DG's with the technological advancements the meta heuristic approaches were utilized since a lot of advantages are there with these techniques such as lesser time is needed for their execution, the outcomes are more accurate and busses number is more which are likely to be dealt in the same time and while using other analytical approaches it is always considered difficult [72]. Different authors have produced various techniques for the analysis which include PSO which stands for particle swarm optimization, AC which is an abbreviation of Ant Colony, FL means frog leap, SA which stands for simulated annealing etc. Glover was the researcher who first used the term meta heuristic and also he suggested TSA which is a famous algorithm and is also known as Tabu Search algorithm. Since there has been much advancement in the field of technology and thus in the existing methods the disadvantages are easy to overcome [72]. Higher rate of convergence was seen with this suggested technique and hence it is considered as highly effective. Colored images were produced with the help of these techniques and these are used for the extraction of the colored images in the form of optimal clusters [79]. The problems such as economic load dispatch was also solved and for this purpose the flower pollination algorithm was used and in this the objective of declining the cost of the fuel is estimated and it is done by the effective real power settling from various generators. [75].

For the solution of problems such as multi objective optimization the SFLA is improved and in this way some global and local explorations are enhanced. They are avoided from getting trapped into the local optima and hence the time of computation decreases as well. The initial population quality is therefore improved. In SFLA the measured enhancements are used as well and they are generally based on the statistical outcomes keeping in mind the modifications which are more common and highly effective. A lot of research workers have recommended these modifications. In the end, the SFLA is addressed making use of quantitative validations such as algorithmic robust in common applications and optimization algorithms are outperformed by it [109]. The optimization problems of these types are used for the improvement of performance and also it is suggested by the previous work that modified SFLA and MSFLA need modifications.

One more step has been considered by the research workers into the MSFLA suggested and in this step, number of frogs is chosen for each memeplex on the basis of their higher value of fitness and sub memeplex are made from the chosen ones and the independent local explorations are made from it. The rate of convergence and the time of processing could be improved via sub memeplex.

1.7.4 Hybrid MSFLA-PSO method

In the algorithms existing, the enhancements can also be brought by the combination of various benefits of other existing algorithms. Hybrid algorithm is a name given to such techniques and for the basic improvement of algorithms the advanced versions are used and this is done to improve the rate of convergence and search space. The effectiveness of individual basic algorithms is the main goal of hybridization and also local explorations are improved by the search space expanding and convergence enhancement. To handle continuous and multiple objective problems of optimization, it is likely to design the flexible, coherent as well as effective algorithm. In the past few years, for HSFLA structures were suggested and made by the research workers in the past. And the suggested algorithms are highlighted in this section and the focus is kept on their specific benefits.

The population based SFLA algorithm is the one which stimulates the frog's behavior which are located in swards looking for food and their optimal locations. The local and global search capabilities are used in this algorithm [119] and MA and PSO algorithms are integrated in it as well. MA has some advantages as well; the biggest advantage is that among all the population the information is passed while in case of GA, another algorithm the information is passed only among parents and siblings. However, faster rate of convergence is there in PSO algorithms and particles are there as well which tend to converge the best worldwide outcomes and solutions and offer a solution with higher speed of convergence and has a real optimal solution [105]. Along with this, a simple process of coding is needed in MA. Therefore, there are many benefits of using these two algorithms named MA and PSO for the profits and advantages in MSFLA. The rate of global search is high and hence it can produce optimal solutions.

SFLA population also has some other approaches such as meta heuristic memetic approach which contains solutions (frogs) and into various memeplexes they are classified on the basis of variety of cultures in frogs. Local exploration is done generally by SFLA along with another PSO algorithm simultaneously in each memeplex. To ensure the global or worldwide explorations the frogs are shuffled and recognized [120].

For the improvement of SFLA and its potential regarding global searching [121] another method is employed which is known as MSFLA. By the modification of SFLA division method this algorithm was suggested and memeplex performance is balanced as well along with the division of SFLA. With the help of new rule of frog leaping, more chances are given for the frogs and in this way, there is an evolution of the best frog to be. MSFLA is much better as shown in the outcome of this investigation and as compared to other MSFLA and PSO algorithms it is considered better and superior. Then the research is carried out to show the improvement in the level of accuracy and optimization with better performance is obtained with modified shuffled frog leap in combination with particle swarm optimization technique for hybrid operation [121]. The observations concluded that the MSFLA-PSO technique outperformed the other techniques by improving the accuracy and speed of operation by guiding the search toward a feasible solution.

1.8 Motivation

The main aim of power system is to provide a sustained power quality for all the users. Initially the power systems were designed for power flow in only one direction from small number of generators to wide areas of demand through transmission and distribution networks. As the electricity needs to be used as and when it is generated as it cannot be stored in large amounts and the operators have to balance the generation and requirement. The system should be made more reliable and sustainable. There is need for coordination of integrating Distributed Generators (DG's) to meet the system technical demand and economic demand by optimal allocation of distributed generators. Hence the optimal allocation and sizing of DG's find a major application in providing efficiently operating power system using algorithms for the decision making.

The DG is the abbreviation of Distributed Generation, for its installation the rapid increase is needed, especially on the basis of RE (the renewable Energy sources) and it is expected that they will be able to address the concerns regarding environment. DGs which are based on the RE or renewable energy, the mature generators are wind and photovoltaic or PV generators and among all the DG related technologies they are fastest in terms of growth. For the smaller grids the natural extensions are DGs or decentralized generations. There is a better management of the central grid due to their onsite decentralization power generation potential and this also helps to reduce the peak loads. For the power system and its proper optimization, strategies regarding the management of energy are used and this is always important to regulate the DGs output powers. The regulation of voltage and grid is also carried out. Furthermore, in grid for the improvement of power quality, there is a need of proper design, level of penetration, control and DG's location, all these factors are equally important.

Power quality is the major issue to be considered in the power system when large penetration of Renewable DGs is considered and maintaining the rated magnitude and frequency of near sinusoidal rated voltage and current of a power system is the tedious job to be done. Interruption of the power quality would cost the efficiency of the system. In most of the cases, control of the power quality refers to the control of the voltage only. This is because the voltage can be controlled more easily than current. More specifically, the quality of power can be described by some parameters such as continuity of service, variation in voltage magnitude, transient voltages and currents, harmonic content.

The main aim of this research is developing different methods for the integration of renewable DGs with large penetration to solve the problem of power quality of the power system. The power quality issues are analyzed and solved by considering the voltage and power loss profiles with the defined indices. An intelligent algorithm is developed to improve the system performance.

1.9 Objectives

The main objectives of the work are as listed below,

a) To analyze voltage and Power loss profiles for different penetration levels by allocating the renewable distributed generators by formulating a weighted sum of objective function.

b) To develop a modified shuffled frog leap algorithm to analyze the impact of integration using different indices based on power quality and to find optimal location for different penetration levels of renewable DG's.

c) To analyze the impact of different penetration levels and optimal integration of renewable DGs on voltage profiles and active power losses by developing a hybrid algorithm.

The work is done in the "MATLAB platform and the algorithms are tested on the standard IEEE 33, IEEE69 and 32 bus practical systems".

Analytical methods, heuristic and Meta heuristic-based approaches are developed to find optimal location of renewable distributed generators for different large penetration levels in this research work. The results of the performance are compared for different cases namely with and without DG at different penetration levels in order to show the effectiveness of the proposed algorithms for the allocation of DG's with their penetration level.

1.10 Methodology

The research is carried out on the standard and practical test systems which are IEEE 33 bus test system, IEEE 69 test system, and practical 32 bus network, the MATLAB software is used for the network analysis and simulations

- An analytical approach is used for analyzing the power loss and voltage profiles. The load flow for the network is obtained using the load flow technique to calculate Logarithmic voltage deviation index (LVDI) and Voltage profile index (VPI) which are used to find out the weighted sum of voltage profile index (WSVPI) from which the location of DG will be determined. Then DG is integrated with the fixed penetration level at the bus selected from the obtained values of WSVPI. Load flow for the network is obtained with the integration of DG to calculate voltage profile improvement index (VPII). The values of VPII will be the evidence to show the impact of DG integration in the network. This method utilizes solar and wind types renewable DG's adaptively for finding the best location of the DG's for fixed penetration level. Integration of solar, wind and hybrid DGs is considered for the analysis.

- The Modified shuffled frog leap algorithm is used for the selection of DG location with respect to different penetration levels of Renewable DGs. On integration of particular type of DG with fixed penetration level power quality-based indices are evaluated to show the improvement. Initially the system is evaluated under normal condition with the integration renewable DGs then the system is evaluated for heavy load condition.

- A Hybrid algorithm based on shuffled frog leap and particle swam optimization is used for the analysis on voltage profiles and active power losses for DG integration at different penetration levels and optimal allocation.

- The code for the algorithm is written in MATLAB

- The results obtained from the analysis of allocated DG and the voltage profile enhancement and power losses reduction are compared using different types of DG's and loading conditions. From there conclusions and scope for future work were discussed.

1.11 Outline of the dissertation

The thesis is organized as follows

Chapter 1 gives the introduction to the power system, distributed generation and Renewable distributed generation, the details of the dissertation giving the review of complete literature surveyed with respect of integration of distributed generators and their effects on the power system.

Chapter 2 gives the analytical approach that computes the weighted sum of voltage profile index and based on the obtained value the location of the DG is determined. That is the candidate bus with the minimum value of the index i.e., the bus with largest deviation is selected for the placement of distributed generator. Through this approach of allocation the voltage profile of the system is improved and hence the results present a better operating system with higher quality of power. This concept has been applied to the IEEE 33 bus feeder system and practical 32 bus system.

Chapter 3 proposes a technique to optimally locate the renewable distributed generators in the considered system for the selected penetration level. In this concern the results are obtained for different indices for penetration of renewable of DG's are obtained using Modified shuffled frog leap algorithm. The technique is testes on IEEE 33 bus system and practical 32 bus system. The application includes the installation of solar and wind powered DG's to achieve enhanced system operation. The simulation results showcase that the develop algorithm reduce the real power loss, improve the voltage profile and quality of the power. The result finally compares the result of this technique with other methods proposed in the literature.

Chapter 4. This chapter proposes a new hybrid method using shuffled frog leap algorithm (SFA) and particle swarm optimization (PSO). To achieve efficient operation, in the first case DG is integrated with different penetration levels and the algorithm gives the best location for the selected penetration level.

Further DG is integrated for optimal allocation using hybrid algorithm. In both the cases solar and wind DGs are used with different cases and scenarios to analyze the voltage profile improvement and to reduce the active power losses for different cases. The proposed technique is applied to the IEEE 33, IEEE69 and practical 32 bus systems. The results in this chapter are compared with the previous chapter.

Chapter 5 presents the conclusions and scope for future work.

1.12 Summary

This chapter describes the general introduction to the electrical power network and the current scenario of the power sector. The distribution network and the need and importance of the Renewable distributed generation are discussed. Also, the note on power quality and the review of the refereed literature are represented. Section 1.5 gives the motivation for the work followed by the details of the dissertation.

CHAPTER 2

2 An analytical method for optimal allocation of Renewable Distributed generation in power system for real power loss reduction and voltage profile improvement

2.1 Introduction

The DG sources have gained attention due to their benefits for increasing the quality of power and more reliable system. But to harness the benefits, it is required to optimally allocate these generators, for which a lot of literature has presented different techniques of placement. In this report indices are defined as decision making factors for allocating DG's.

The main objective of the work is to minimize the real power losses and improve the voltage at the nodes. The research makes the analysis on the IEEE 33 bus test system and 32 bus practical test system and the simulation results are presented for the code written in MATLAB platform.

As the deviation of the voltage is high a farther bus from the feeder the voltage profile at the far ends is not good. This voltage deviation will lead to a poor voltage at the end busses, thereby affecting the quality of the power delivered to the customers. In this research a new algorithm is developed to determine the optimal location at particular penetration level to reduce the power losses and improve the voltage profiles. This algorithm gives the best solution which is based on logarithmic voltage deviation index and power loss reduction simultaneously.

"The location of DG is chosen as the one that gives the best voltage profile. This could be done by injecting the DG with a particular penetration level. The best location is chosen based on the multi objective function obtained from the voltage profile index and logarithmic voltage deviation index."

2.2 Literature review

Distributed generators are connected to the electrical network at their load points on the customer site of the meter [7]. Also, it acts as a backup source during the failure of the main supply also during the unavailability of the power as well as an additional source during the normal operation to reduce the system losses [56]. Integration of DG to the system under operation is an art and important for considering losses in the system, quality of power obtained and the variation of the voltage at different points. [65]. It is a practice to use analytical approach for optimal placement in the initial cases to reduce the total system losses by improving voltage profiles to increase the capacity of the system was proposed by Griffin.

The Rule of thumb is used in the paper [59] which includes two methods as zero-point analysis used at the points where power flow is zero from DG and another method called 2/3 rule from which placement of capacitors in radial distribution system, according to 2/3 rule the size of the capacitor is decided to two thirds of kVAR load when it is located at 2/3 of the distance from the feeder.

In the paper [60] authors have developed an analytical method for multiple DG placement to improve system performance by using indices. In this case, the simple power flow method is being used and generator buses are treated as PV buses. According to authors in [61], DGs can be placed depending on the different type of loads as well as DG sources, an optimal location was chosen based on the bus admittance matrix with respect to generation and distribution of load. The paper [62] suggests the placement of DG based on the effect of load modeling, a classical grid search algorithm is used for each load and the analysis is carried out for constant current, power, and constant impedance models and according to this approach the location of DG will remain the same even if the load got increased.

The economic impacts and time-varying loads will change the system reliability and its efficiency, optimal placement of DG under this condition is proposed by authors in [64]. A method based on a technical and economic assessment on biomass-fueled generators, load curtailment is used for managing the power and DGs are installed based on the sensitivity of losses and cost analysis.

Authors of the paper [66] have used loss sensitivity factor based optimal DG allocation without considering either admittance nor jacobian matrix, here grid flow logarithm is modified based on load flows to minimize losses in the system. Authors in paper [68] have proposed a method to install different types of DGs with respect to different loads based on the consideration of power factor.

Authors in paper [69] have proposed a method of allocating DG based on critical node voltage deviation (CNVDI) where the node with more sensitivity is chosen to obtain optimal location for the installation of DG which is done for different penetration levels to get an improved voltage profile and to reduce the system losses. Authors of the paper [70] have proposed a method of Tail end node voltage deviation (TNVDI) by selecting the end nodes of the test system for choosing a location and different types of DGs are installed in a range to find the optimal size.

2.3 Problem formulation

2.3.1 Penetration level:

The penetration level of DG is chosen based on the following equation.

$$\mathbf{PLDG\%} = \frac{\textbf{Size of DG}}{\textbf{Total demand}} * \mathbf{100} \quad \dots\dots\dots\dots\dots\dots\dots\dots\dots\dots\dots\dots\dots\dots\dots\dots 2.1$$

Where, PLDG% is penetration Level of DG [139]

A PL of 0% represents when the load demand is totally met by the grid and a 100% PL represents when the load demand is supplied entirely by the DG Technologies.

2.3.2 Power loss calculation:

The equation of the real power losses in the system can be given in terms of exact loss formula as state below. The power loss is given by

$$\mathbf{P_L} = \sum_{i=1,j=1}^{n} [\boldsymbol{\alpha}_{ij}(\mathbf{P_i P_j} + \mathbf{Q_i Q_j}) + \boldsymbol{\beta}_{ij}(\mathbf{Q_i P_j} - \mathbf{P_i Q_j})] \dots\dots\dots\dots\dots\dots\dots\dots\dots\dots\dots 2.2$$

Where $\boldsymbol{\alpha}_{ij} = \frac{r_{ij}}{V_i V_j} \mathbf{Cos}(\partial_i - \partial_j)$ and $\boldsymbol{\beta}_{ij} = \frac{r_{ij}}{V_i V_j} \mathbf{Sin}(\partial_i - \partial_j)$ [139]

$$\mathbf{Q_L} = \sum_{i=1,j=1}^{n} [\mathbf{¥}_{ij}(\mathbf{P_i P_j} + \mathbf{Q_i Q_j}) + \boldsymbol{\lambda}_{ij}(\mathbf{Q_i P_j} - \mathbf{P_i Q_j})] \dots\dots\dots\dots\dots\dots\dots\dots\dots\dots 2.3$$

Where $\mathbf{¥}_{ij} = \frac{x_{ij}}{V_i V_j} \mathbf{Cos}(\partial_i - \partial_j)$ and $\boldsymbol{\lambda}_{ij} = \frac{x_{ij}}{V_i V_j} \mathbf{Sin}(\partial_i - \partial_j)$ [139]

$$\mathbf{WSVPI} = ((\mathbf{w1} * \mathbf{VPI}) + (\mathbf{w2} * \mathbf{LVDI})) \dots\dots\dots\dots\dots\dots\dots\dots\dots\dots\dots\dots\dots 2.4$$

Where,

WSVPI= weighted sum of voltage profile index

Where, the sum of weights (w1+w2) must be equal to unity. The values of w1 and w2 must be chosen in such a way that they must sum up to 1. [preferably 0.5 each].

Voltage profile index (VPI) is given by

$$VPI = V_{ref} - V_i \dots\dots\dots\dots\dots\dots\dots\dots\dots\dots\dots\dots\dots\dots\dots\dots\dots\dots 2.5$$

where,

In this case, the magnitude of the voltage of bus is Vi the bus slack voltage magnitude is Vref for VPI opportunity is offered by it to aggregate as well as quantify the level of voltage, importance and its amount where there is a supply of load in the system of load busses.

The analysis is carried out in such a way that the voltage profile index of the system is 0 at 0.95 pu (Vmin) and at 1.05 pu (Vmax) and has a maximum value of 1.0 when all the bus voltages are at their nominal (desirable) values. If any bus voltage falls below or rises above the nominal voltage, the overall voltage profile of the system is affected adversely, thus providing a true picture of the effect of DG on the overall system. VPI of individual nodes will be negative when bus voltage falls below 0.95 or rises above 1.05, leading to a reduction of the overall voltage profile index of the system

2.3.3   Voltage Profile Improvement Index:

At various buses there is an addition of DGs and as a result voltage profile is improved. The VPII index is there for the improvement of voltage profile and this also quantifies the voltage profile or VP improvement and this is done by the addition of DG. The following equation is used to express it.

$$VPII = \frac{VPI\ w/DG}{VPI\ wo/DG} \dots\dots\dots\dots\dots\dots\dots\dots\dots\dots\dots\dots\dots\dots\dots\dots\dots\dots 2.6$$

The following attributes Based on this definition are:

VPII < 1, DG for the system has better and improved voltage.

VPII = 1, on the system voltage profile DG has no impact.

VPII > 1 no beneficial impact is there for DG.

Where, VPI $_{w/}$ DG, VPI $_{wo/}$ DG are used to represent the system's voltage profile measurement respectively with and without DGs.

2.3.4 Logarithmic Voltage Deviation Index

$$LVDI = \sum_{i=1}^{n} log(\frac{V_{i(new)}}{V_{i(old)}}) \quad where\ i = 1,2,3 \dots \dots n \dots \dots \dots \dots \dots \dots \dots 2.7$$

Where, V_{old} is the voltage obtained after performing the load flow and V_{new} is the value of the voltage after placement of DG.

Summation of all two voltage deviations with the multiplication of corresponding weighing factors should be equal to unity. In this work, the DG is placed at the bus with minimum value of WSVPI.

2.3.5 Modeling of WIND and SOLAR DGs

The Photovoltaic systems convert Solar Energy into Electrical energy. Their output is DC power and is converted into AC power via an inverter to be compatible with AC grid. The power output of a solar photovoltaic (PV) panel depends on the area (A) of the PV panel, solar irradiance $\mu(t)$ and efficiency of the PV panel β[18]

P pv(t) = A β μ(t) $\dots\dots\dots\dots\dots\dots\dots\dots\dots\dots\dots\dots\dots\dots\dots\dots$2.8

The Wind turbine power output[18] is proportional to the kinetic energy, air density, etc. Other parameters of Wind turbine include cut- in wind speed, cut-out wind speed and rated wind speed, and typical values of them are 3.5 m/s, 25 m/s, and 14 m/s respectively.

P_{wind} (t) = 0.5 α ρ(t)A v(t)^3$\dots\dots\dots\dots\dots\dots\dots\dots\dots\dots\dots\dots$2.9

Where α is the Albert Betz constant, ρ(t) is air density, A is area swept by turbine rotor, and v(t) is wind speed [18].

The steps of execution of the algorithm are given as follows

Step1: Select the standard test system with all base case parameters

Step 2: Run load flow on the selected test system without integration of DG

Step 3: select obtained values and calculate VPI and LVDI.

Step 4: Calculate WSVPI using equation 2.4

Step 5: Select the bus giving maximum value of WSVPI which is considered as optimal location.

Step 6 Integrate the DG at the optimal location and run load flow (with different penetration levels as 15%,20%,30%)

Step 7: Calculate the power losses and VPII obtain the voltage profiles and their improvement.

In this work, the DG is placed at the maximum WSVPI nodes and that gives the optimal location. Initially DG is integrated by choosing VPI and LVDI method and corresponding values of each method are obtained and multiplied with weighing factor to get the WSVPI values at each node.

Then DG with particular penetration level is integrated at the maximum WSVPI node to obtain the better results. The improvement of voltage profiles with the integration of DGs is analyzed with the values obtained from the Voltage profile improvement index.

2.4   PROPOSED ALGORITHM

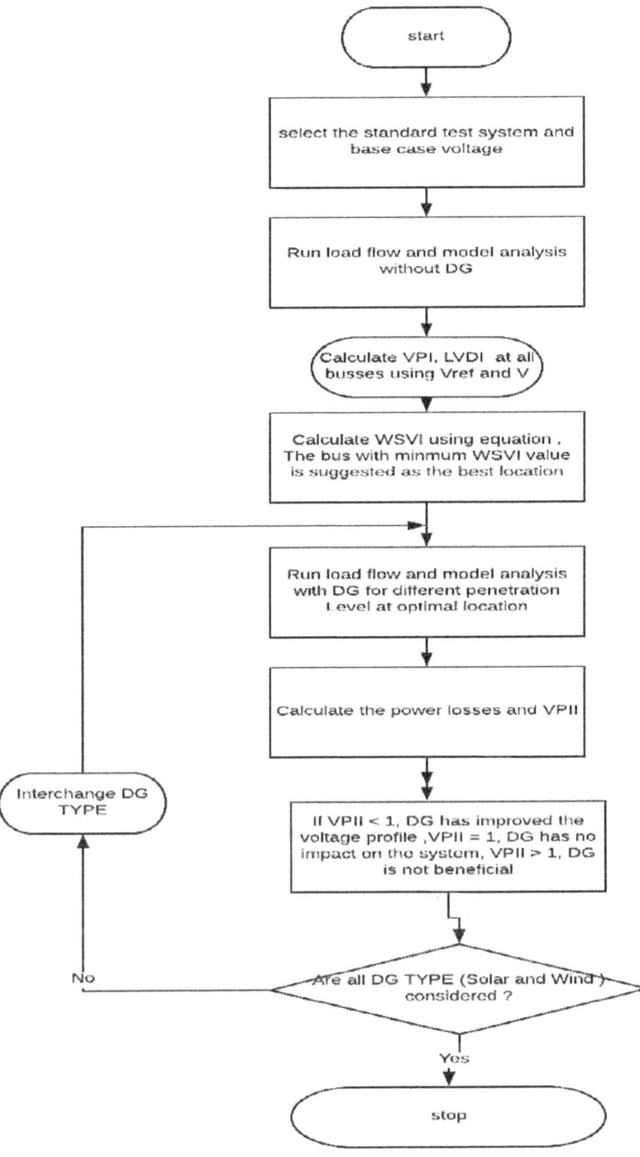

Fig 2.0 : Flowchart of the proposed algorithm

33

2.5 Results and discussions

The proposed technique is tested on the standard IEEE 33 bus system and practical 32 bus test system in MATLAB platform. The results obtained are discussed.

2.5.1 Case Study 1: IEEE 33 Bus test system

The proposed method is tested on the standard IEEE 33 bus system operated at 12.66kV level of voltage and consists of 33 busses 32 branches. 3715kW and 2300kVar is the total active and reactive power of the system respectively.

Fig 2.1 shows the diagram of the network considered. The basic data of the network that is the line data and the branch data are given in appendix A. The nodes with the highest amount of deviation are identified and tabulated in the table. The base case load flow is obtained where the power losses without the penetration of the renewable DG's are obtained to be 202.68kW and 143.22kVAr respectively.

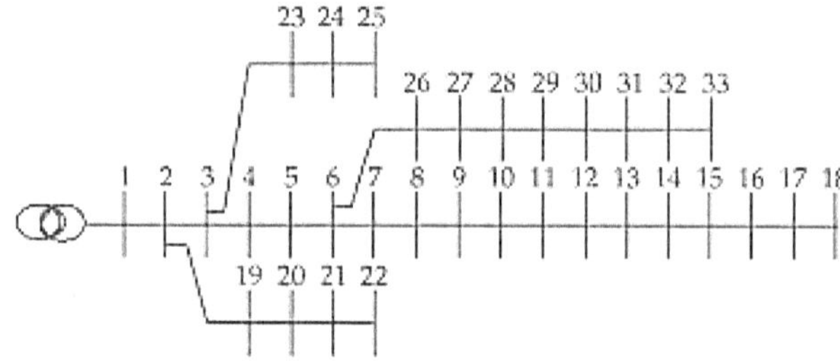

Figure 2.1: IEEE33 bus test system

Fig.2.2 shows the voltage profiles before and after DG placement at different penetration levels say 15, 20 and 30%. The variation of voltage at different buses of the 33-bus test system can be observed in the figure2.2 and the voltage profile has been improved with increase in the penetration level. Fig.2.3 shows the system active power losses before and after placement of solar DG at different penetration levels.

Fig. 2.4 shows the Values of voltage profile improvement index at different buses of the system with solar DG penetration at different levels.

The Base Case represents the scenario in which no DG was connected to the network is to allow us make a good comparison with when DGs are connected.

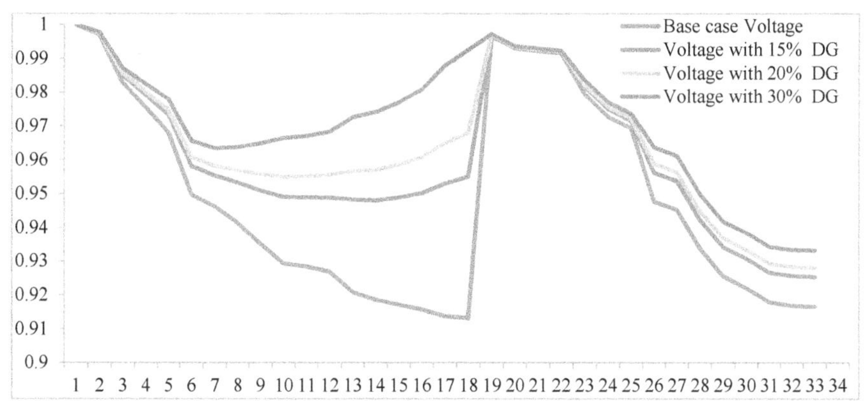

Fig.2.2: Voltage profiles of the system with solar DG penetration at different levels

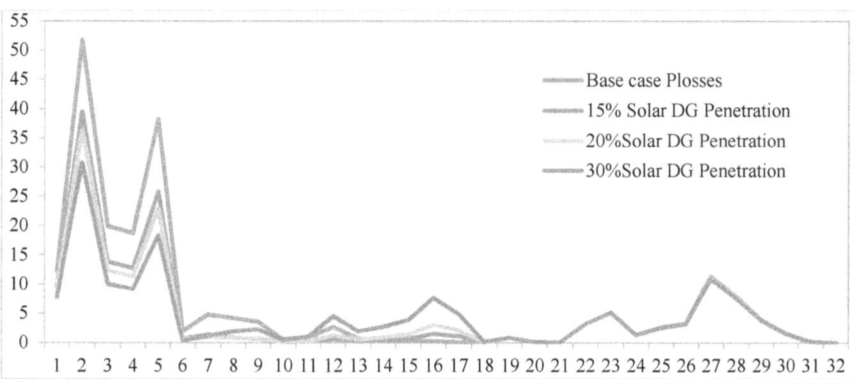

Fig.2.3: Active power loss variation at different buses of the system with solar DG penetration at different levels

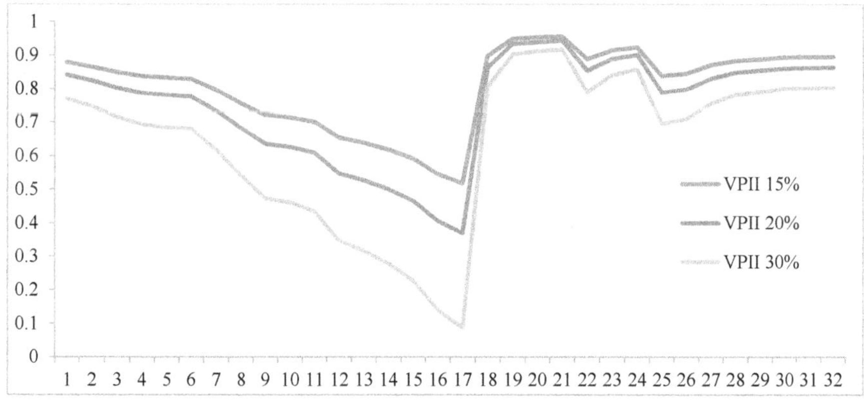

Fig.2.4: Values of voltage profile improvement index at different buses of the system with solar DG penetration at different levels

35

Table 2.1: LVDI, WSVPI for different penetration levels.

Solar DG at different penetration levels

Bus no	At 15% DG Penetration		At 20% DG Penetration		At 30% DG Penetration	
	LVDI	WSVPI	LVDI	WSVPI	LVDI	WSVPI
1	0	0	0	0	0	0
2	0.000157	0.001383	0.000205	0.001352	0.000296	0.001293
3	0.001011	0.00789	0.001319	0.007695	0.001905	0.007322
4	0.001649	0.011239	0.002158	0.010919	0.00311	0.01032
5	0.002327	0.014533	0.00304	0.01409	0.004388	0.013249
6	0.003847	0.022869	0.00502	0.02216	0.007216	0.020823
7	0.004207	0.024413	0.005478	0.023649	0.007846	0.022213
8	0.00546	0.02611	0.007111	0.025121	0.010204	0.023247
9	0.007282	0.028206	0.009487	0.026888	0.013602	0.024396
10	0.009143	0.030066	0.011903	0.028421	0.017065	0.025292
11	0.00949	0.0303	0.012355	0.028592	0.017714	0.025342
12	0.010151	0.030676	0.013216	0.028848	0.018951	0.025365
13	0.012749	0.032274	0.016593	0.029981	0.02377	0.0256
14	0.013715	0.032873	0.017842	0.030411	0.025547	0.025698
15	0.01474	0.032995	0.019171	0.030346	0.027443	0.025266
16	0.016025	0.032942	0.020839	0.030054	0.029822	0.024506
17	0.01823	0.03268	0.023701	0.029376	0.033896	0.023013
18	0.019467	0.032259	0.025308	0.038714	0.036186	0.021878
19	0.000157	0.001648	0.000209	0.001615	0.000296	0.001558
20	0.000157	0.003434	0.000206	0.003403	0.000297	0.003344
21	0.000158	0.003789	0.00021	0.003755	0.000302	0.003696
22	0.000162	0.004106	0.00021	0.004075	0.000302	0.004016
23	0.001019	0.009684	0.001333	0.009486	0.00192	0.009115
24	0.001035	0.013017	0.001351	0.01282	0.001947	0.012448

25	0.001038	0.014679	0.001355	0.014483	0.001958	0.014109
26	0.003864	0.023832	0.005039	0.023124	0.007244	0.021787
27	0.003884	0.025112	0.005066	0.024403	0.007281	0.023066
28	0.003982	0.030826	0.005192	0.030116	0.007466	0.028773
29	0.004054	0.034932	0.005284	0.034222	0.007596	0.032878
30	0.004088	0.036709	0.005328	0.035999	0.007657	0.034654
31	0.004121	0.03879	0.005375	0.038078	0.007724	0.036732
32	0.004134	0.039247	0.005385	0.038538	0.007741	0.037191
33	0.004136	0.039388	0.005387	0.038678	0.007744	0.037332

Table 2.1 shows the values of Logarithmic Voltage Deviation Index and Weighted Sum of Voltage Profile Index for the penetration levels of 15, 20 and 30 percent respectively. Here the optimal location for solar DG penetration at particular penetration level is chosen based on the maximum value of weighted sum of voltage profile index values obtained from all the buses of the system.

It can be clearly observed in the table that the values of Logarithmic Voltage Deviation Index and Weighted Sum of Voltage Profile Index are less than unity at all busses of the test system and this the evident that the integration of DG has improved the system performance by improving the voltage.

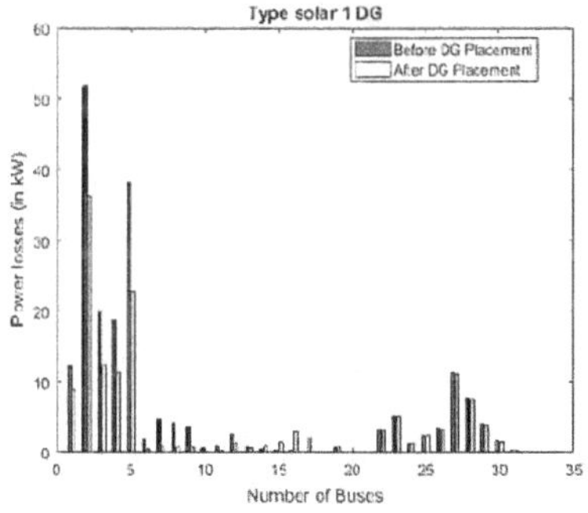

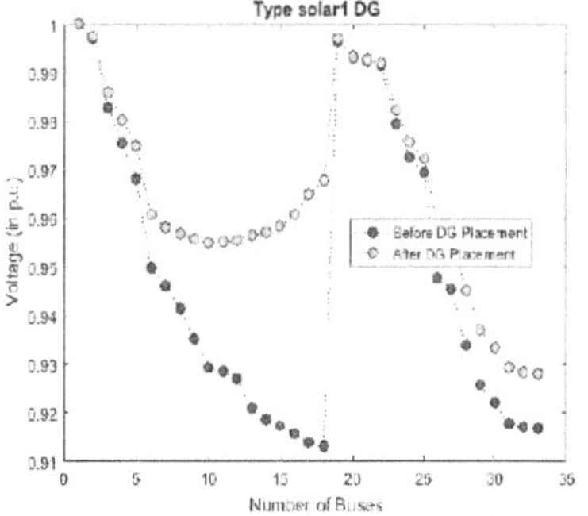

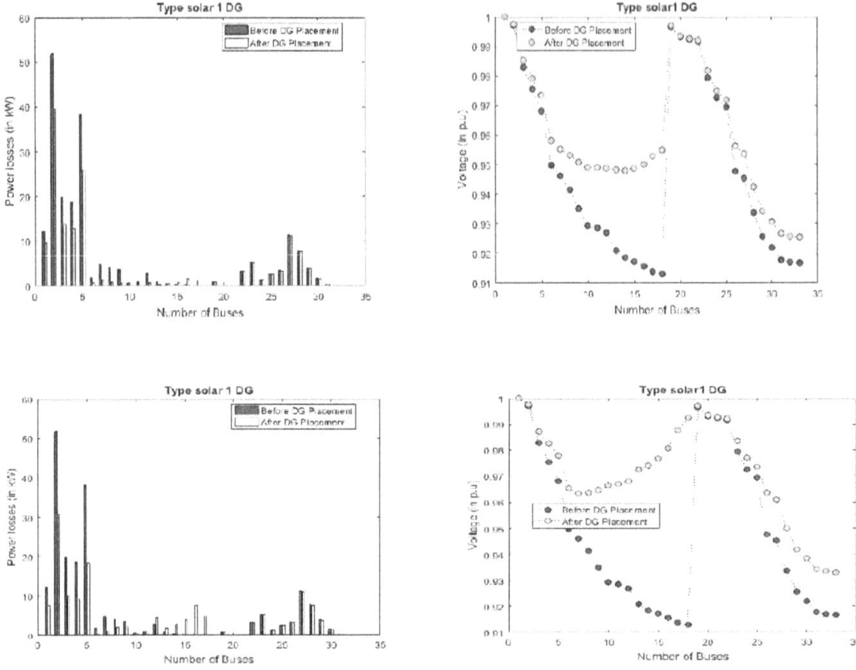

Fig.2.5: Voltage profile and power losses of IEEE33 bus system for integration of solar DG at 15%, 20% and 30% Penetration

Figure 2.5 shows the plot of voltage profile improvement index for solar DG penetration at different levels

The results obtained from the simulation are shown in figure 2.6. The variation of voltage profiles and power losses at different buses of IEEE 33bus test system for integration of wind powered DG with 15, 20 and 30% penetration.

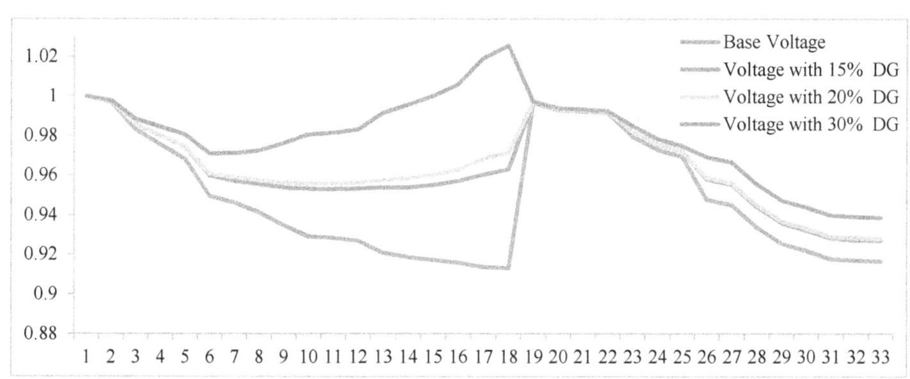

Fig.2.6: Voltage profiles of the system with Wind powered DG penetration at different levels

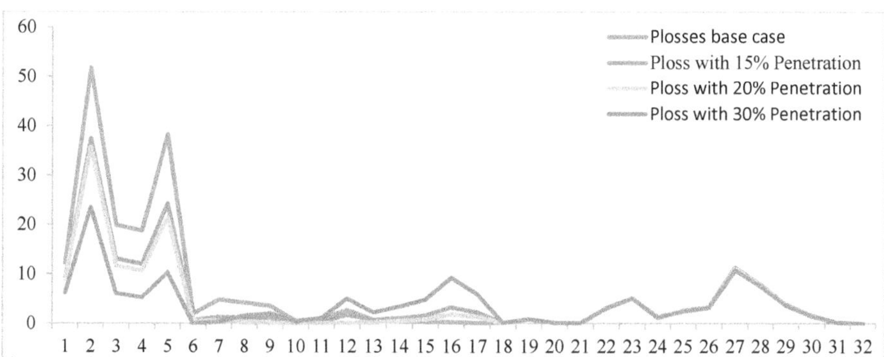

Fig.2.7: Active power loss variation at different buses of the system with wind powered DG penetration at different levels

Fig.2.6 shows the voltage profiles after DG placement at different penetration levels say 15, 20 and 30%. The variation of voltage at different buses of the 33-bus test system can be observed in the figure and the voltage profile has been improved with increase in the penetration level. Fig.2.7 shows the system active power losses after placement of wind DG at different penetration levels.

Table 2.2: LVDI and WSVPI for different penetration levels. .

Bus no	Wind DG at different penetration levels					
	At 15% DG Penetration		At 20% DG Penetration		At 30% DG Penetration	
	LVDI	WSVPI	LVDI	WSVPI	LVDI	WSVPI
1	0	0	0	0	0	0
2	0.000196	0.001263	0.0002	0.001282	0.000379	0.001239
3	0.00124	0.001358	0.001271	0.001375	0.002428	0.006989
4	0.002025	0.007745	0.002074	0.007725	0.003967	0.009778

5	0.002862	0.011003	0.00292	0.010972	0.005581	0.0125
6	0.004653	0.014201	0.005051	0.014165	0.009647	0.019328
7	0.004961	0.022382	0.005926	0.022141	0.011331	0.020075
8	0.006534	0.02396	0.007379	0.023378	0.014072	0.020871
9	0.008723	0.025467	0.009823	0.024959	0.018686	0.021258
10	0.01096	0.027346	0.012303	0.026686	0.023337	0.021399
11	0.011394	0.028985	0.012705	0.028181	0.024085	0.021382
12	0.012224	0.029167	0.013462	0.028382	0.025503	0.021282
13	0.015329	0.029442	0.01701	0.028701	0.032115	0.020338
14	0.01641	0.030739	0.018554	0.02973	0.035001	0.019706
15	0.017619	0.03127	0.019995	0.029982	0.037675	0.018737
16	0.019167	0.031279	0.021706	0.029847	0.040835	0.017418
17	0.021639	0.031064	0.025251	0.029528	0.047413	0.014157
18	0.023126	0.039063	0.026911	0.028426	0.050422	0.012461
19	0.000196	0.030048	0.0002	0.027726	0.000379	0.001504
20	0.000192	0.001623	0.000197	0.00162	0.00038	0.00329
21	0.000197	0.003411	0.000201	0.003408	0.000385	0.003643
22	0.000197	0.003763	0.000201	0.003761	0.000385	0.003963
23	0.001253	0.004084	0.00128	0.004081	0.00245	0.00878
24	0.001266	0.009537	0.001297	0.00952	0.002484	0.012112
25	0.001275	0.012873	0.001306	0.012854	0.002497	0.013774
26	0.004672	0.014532	0.005075	0.014513	0.009684	0.020292
27	0.004698	0.023346	0.005098	0.023103	0.009737	0.021568
28	0.004815	0.024624	0.005229	0.024384	0.009978	0.027274
29	0.004904	0.030338	0.005321	0.030094	0.010152	0.036376
30	0.004941	0.034442	0.005365	0.034201	0.010232	0.033151
31	0.004982	0.036221	0.005413	0.035978	0.01032	0.03523
32	0.004997	0.038301	0.005423	0.038056	0.010344	0.035687
33	0.004998	0.038758	0.005424	0.038516	0.010347	0.035829

Table 2.2 shows the values of Logarithmic Voltage Deviation Index and Weighted Sum of Voltage Profile Index for the penetration levels of 15, 20 and 30 percent respectively. Here the optimal location for wind powered DG penetration at particular penetration level is chosen based on the maximum value of weighted sum of voltage profile index values obtained from all the buses of the system.

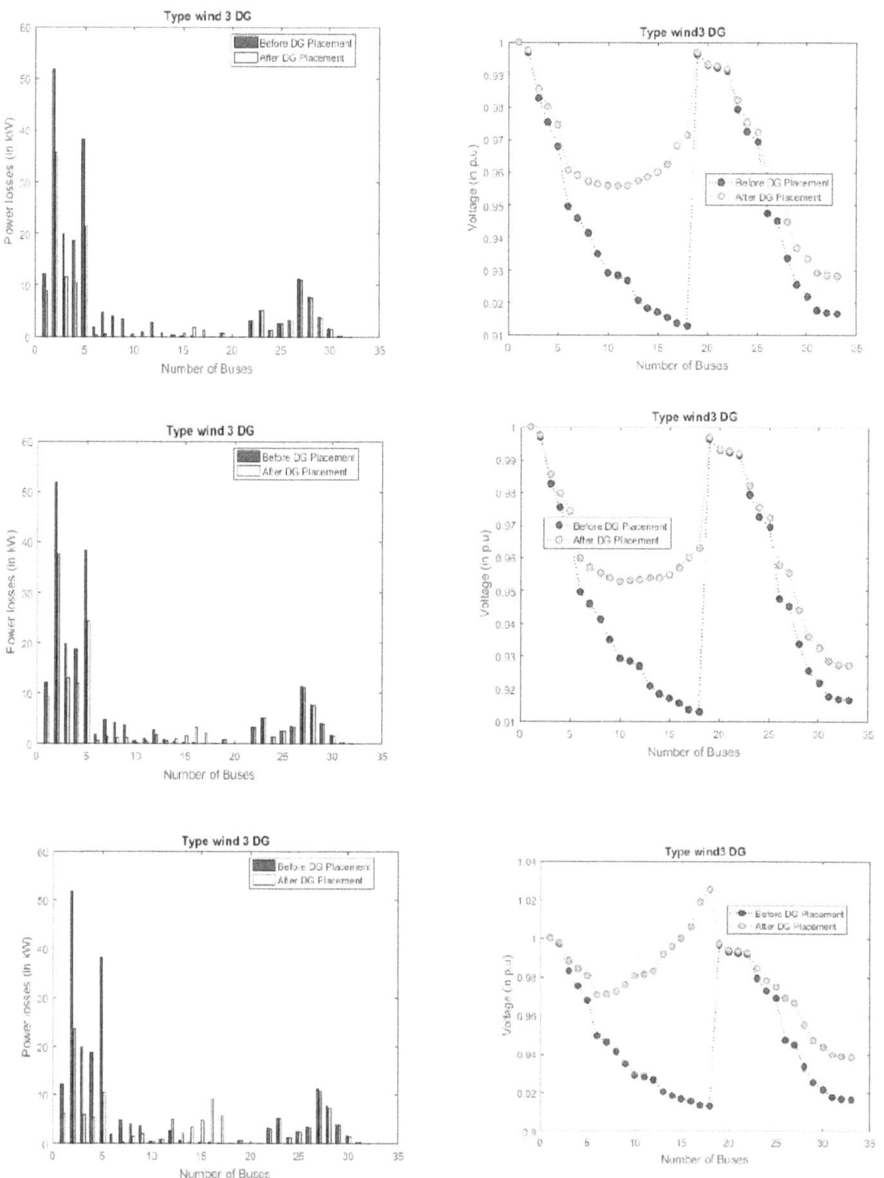

Fig.2.8: Voltage profile and power losses of IEEE33 bus system for integration of wind DG

The results obtained from the simulation are shown in figure 2.9. the variation of voltage profiles and power losses at different buses of IEEE 33 bus test system for integration of wind powered DG with 15, 20 and 30% penetration.

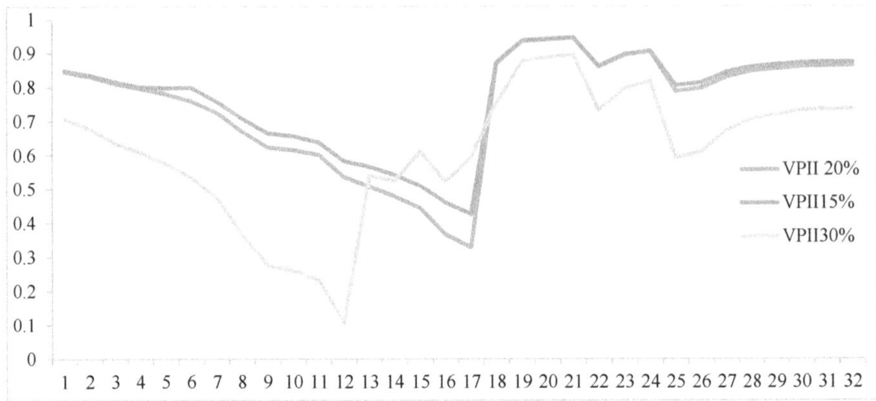

Fig.2.9: Values of voltage profile improvement index at different buses of the system with wind powered DG penetration at different levels

Figure 2.9 shows the different values of voltage profile improvement index at all the busses of the test system under study. It is evident that the system voltage profile has improved with the integration of DG as the values are less than unity at all the busses.

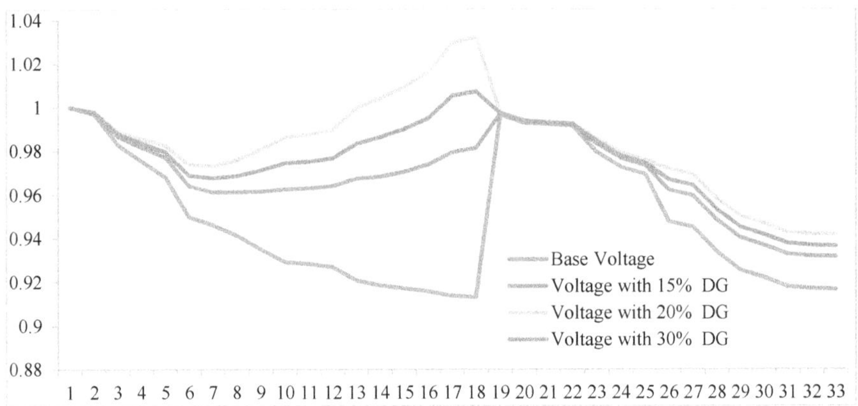

Fig.2.10: Voltage profiles of the system with solar and Wind powered DGs penetration at different levels

Fig.2.10 shows the voltage profiles before and after DG placement at different penetration levels say 15, 20 and 30%. The variation of voltage at different buses of the 33-bus test system can be observed in the figure and the voltage profile has been improved with increase in the penetration level. Fig.2.11 shows the system active power losses before and after placement of solar and wind powered DGs at different penetration levels.

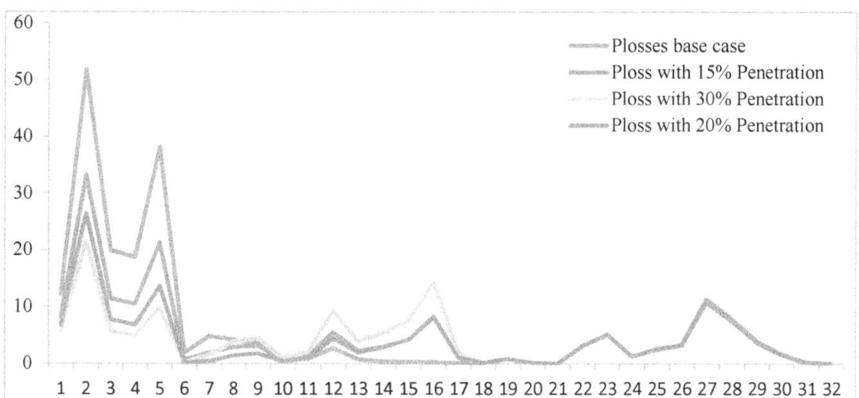

Fig.2.11: Active power loss variation at different buses of the system with wind powered DG penetration at different levels

Table 2.3: LVDI and WSVPI for different penetration levels.

Solar and Wind Hybrid DG at different penetration levels

Bus no	At 15% DG Penetration		At 20% DG Penetration		At 30% DG Penetration	
	LVDI	WSVPI	LVDI	WSVPI	LVDI	WSVPI
1	0	0	0	0	0	0
2	0.000274	0.001622	0.00044	0.001705	0.000348	0.001659
3	0.001755	0.009407	0.002805	0.009933	0.002226	0.009643
4	0.002871	0.013706	0.004584	0.014562	0.003631	0.014086
5	0.00405	0.017995	0.006448	0.019194	0.005115	0.018528
6	0.006536	0.028438	0.01087	0.030605	0.008635	0.029487
7	0.006876	0.030353	0.012318	0.033074	0.009785	0.031808
8	0.009126	0.033898	0.015614	0.037142	0.012412	0.035541
9	0.012191	0.038566	0.02075	0.042845	0.016509	0.040725
10	0.015322	0.043041	0.025934	0.048347	0.020658	0.045709
11	0.015946	0.043783	0.026829	0.049224	0.021372	0.046496
12	0.01713	0.045125	0.028528	0.050824	0.022738	0.047929
13	0.021459	0.050345	0.035805	0.057518	0.02857	0.0539
14	0.022931	0.052215	0.038741	0.060121	0.030934	0.056217

15	0.024615	0.053762	0.041651	0.062281	0.033263	0.058086
16	0.026787	0.055534	0.045174	0.064727	0.036087	0.060183
17	0.030179	0.058239	0.051865	0.069082	0.041492	0.063896
18	0.031368	0.059139	0.053292	0.070101	0.042645	0.064778
19	0.000274	0.001887	0.00044	0.00197	0.000349	0.001924
20	0.000275	0.003673	0.000442	0.003756	0.00035	0.00371
21	0.000276	0.004028	0.000442	0.004111	0.00035	0.004065
22	0.000276	0.004348	0.000442	0.004431	0.00035	0.004385
23	0.00177	0.01121	0.002829	0.011739	0.002243	0.011446
24	0.001796	0.014558	0.00287	0.015095	0.002276	0.014798
25	0.001806	0.016223	0.002885	0.016762	0.002288	0.016464
26	0.006563	0.029416	0.010915	0.031592	0.00867	0.03047
27	0.006598	0.030714	0.010975	0.032903	0.008716	0.031773
28	0.006761	0.036516	0.011244	0.038757	0.008931	0.0376
29	0.006881	0.040685	0.011443	0.042967	0.009092	0.041791
30	0.006935	0.042492	0.011528	0.074789	0.009159	0.043605
31	0.006999	0.044604	0.01163	0.04692	0.009242	0.045726
32	0.00701	0.04507	0.011656	0.047393	0.00926	0.046195
33	0.007017	0.045214	0.011664	0.047537	0.009268	0.046339

Table 2.3 shows the values of Logarithmic Voltage Deviation Index and Weighted Sum of Voltage Profile Index for the penetration levels of 15, 20 and 30 percent respectively. Here the optimal location for solar and wind powered DGs penetration at particular penetration level is chosen based on the maximum value of weighted sum of voltage profile index values obtained from all the buses of the system.

It can be clearly observed in the table that the values of Logarithmic Voltage Deviation Index and Weighted Sum of Voltage Profile Index are less than unity at all busses of the test system and this is evident that the integration of DG has improved the system performance by improving the voltage.

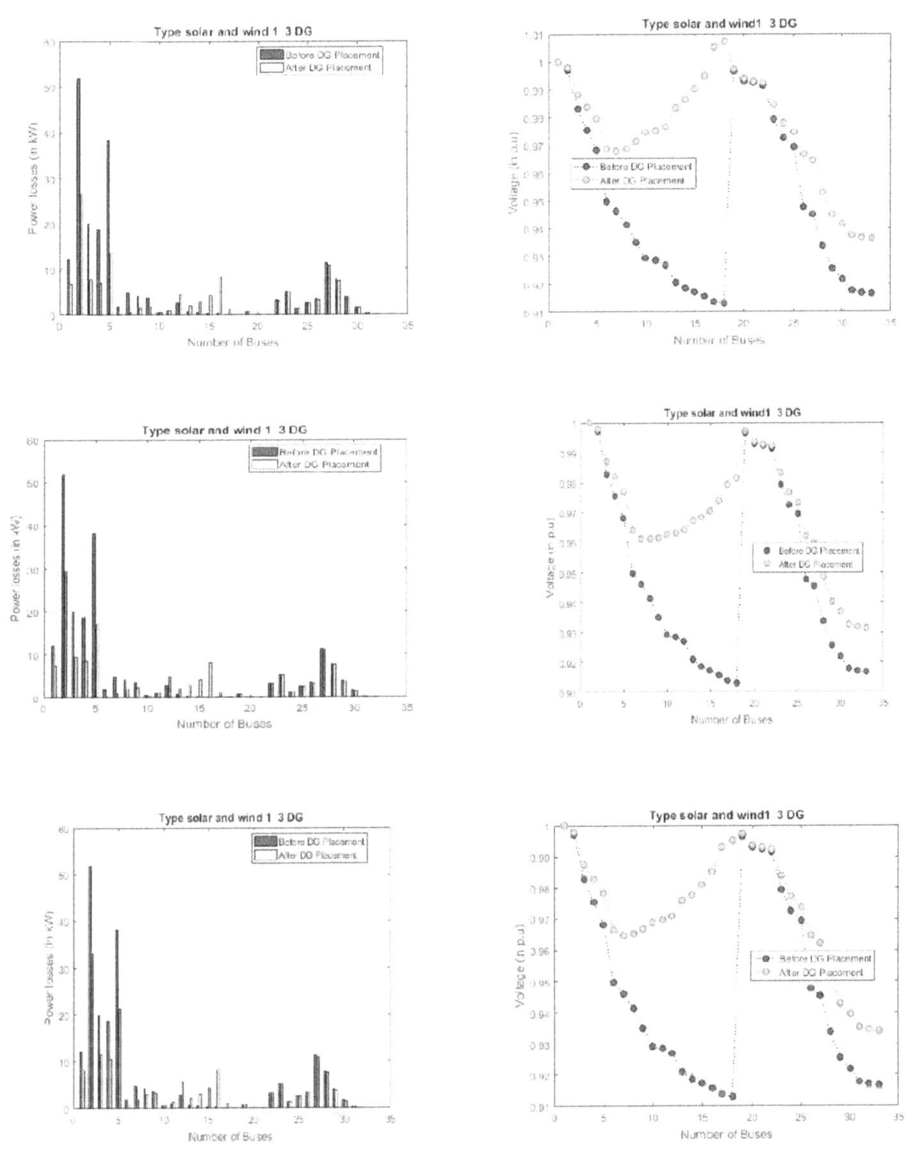

Fig.2.12: Voltage profile and power losses of IEEE33 bus system for integration of solar and wind powered DGs at 15%, 20% and 30% Penetration

The results obtained from the simulation are shown in figure 2.12. The variation of voltage profiles and power losses at different buses of IEEE 33bus test system for integration of solar and wind powered DGs with 15, 20 and 30% penetration.

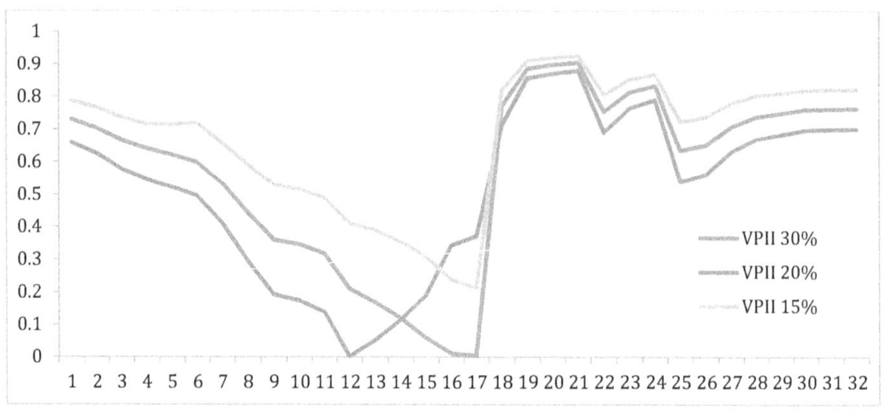

Fig.2.13: Values of voltage profile improvement index at different buses of the system with wind powered DG penetration at different levels

Figure 2.13 shows the different values of voltage profile improvement index at all the busses of the test system under study. It is evident that the system voltage profile has improved with the integration of DG as the values are less than unity at all the busses.

Table 2.4: Summary of voltages and losses for 33 bus system

DG Penetration Level	Optimal Location	DG type	Active power Losses			Minimum Voltage in p.u	
			Without DG (kW)	With DG (kW)	%PLR	Without DG(p.u)	With DG(p.u)
	33	Solar		159.63	21.24		0.92536
15 %	18	Wind	202.68	149.04	26.46	0.9137(18)	0.93152
	33 18	Solar + wind		145.62	28.15		0.94582
	18	Solar		141.54	26.24		0.92803
20%	33	Wind	202.68	142.84	29.52	0.9137(18)	0.93663
	30 33	Solar + wind		129.47	36.11		0.95126
30 %	33	Solar	202.68	134.78	33.49	0.9137(18)	0.9338

| 29 | Wind | 123.71 | 38.96 | 0.94154 |
| 18 29 | Solar + wind | 110.66 | 45.40 | 0.95992 |

Table 2.4 shows the summary of DG integration at different penetration levels at 15%, 20% and 30% respectively. Here renewable DGs such as solar, wind and hybrid solar with wind distributed generators are used to improve the system performance by improving the voltage and reducing the losses.

The losses are reduced the maximum value with solar and wind power hybrid combination for the same penetration level. The losses with the integration of wind powered distributed generation are obtained to be less compared to solar distributed generators, as the solar is capable of generating only active power whereas wind is capable of compensating the reactive power in addition to the real power generation.

2.5.2   Case Study 2: 32 bus practical radial distribution network

The proposed method is tested on the 32 bus radial distribution network (Bhuvaneshwari feeder, mysuru, India) a practical system. The loads are assumed to be constant with the voltage level of 11kV and 20 MVA Base. 3715kW and 2300kVar is the total active and reactive power of the system respectively.

Fig 2.14 shows the diagram of the network considered. The basic data of the network that is the line data and the branch data are given in appendix A. T

he nodes with the highest amount of deviation are identified and tabulated in the table. The base case load flow is obtained where the power losses without the penetration of the renewable DG's are obtained to be 1.1109kW. The optimal location of the renewable DGs for different penetration level is calculated using the WSVPI.

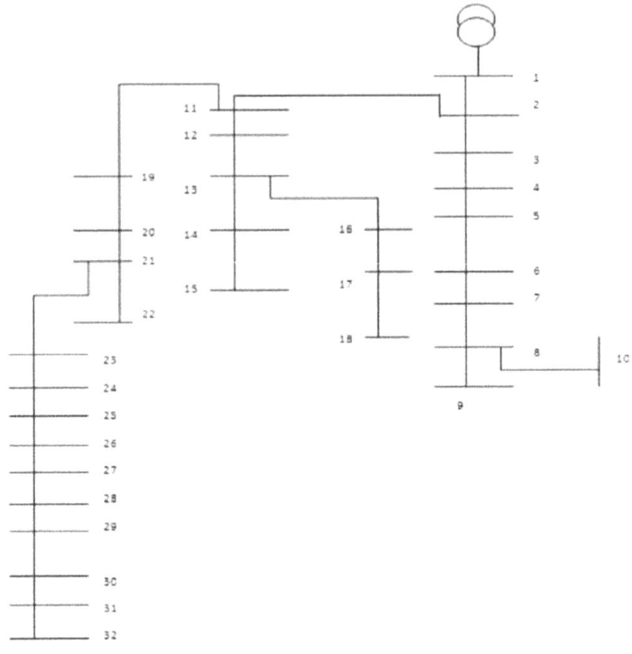

Fig 2.14. Single line diagram of 32 bus practical test system.

Fig.2.15 shows the voltage profiles before and after DG placement at different penetration levels say 15, 20 and 30%. The variation of voltage at different buses of the 32-bus practical test system can be observed in the figure and the voltage profile has been improved with increase in the penetration level.

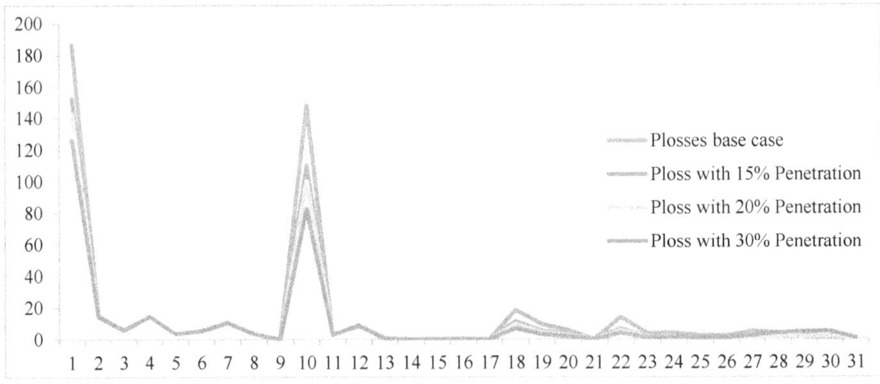

Fig.2.15: Active power loss variation at different buses of the system with solar DG penetration at different levels

48

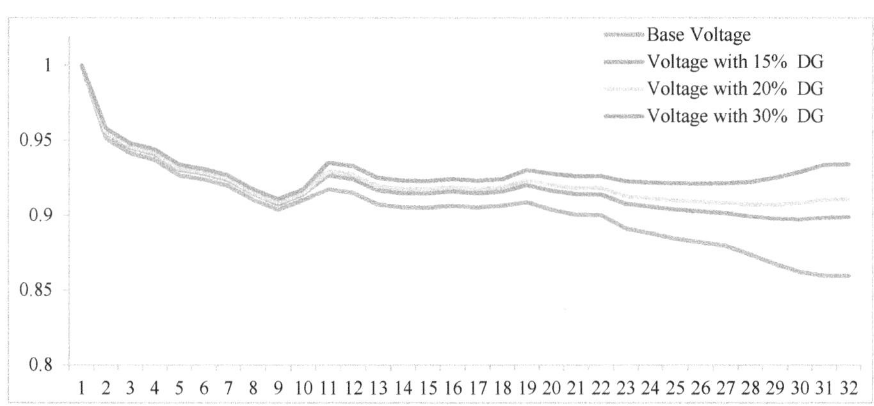

Fig.2.16: Values of voltage profile improvement index at different buses of the system with solar DG penetration at different levels

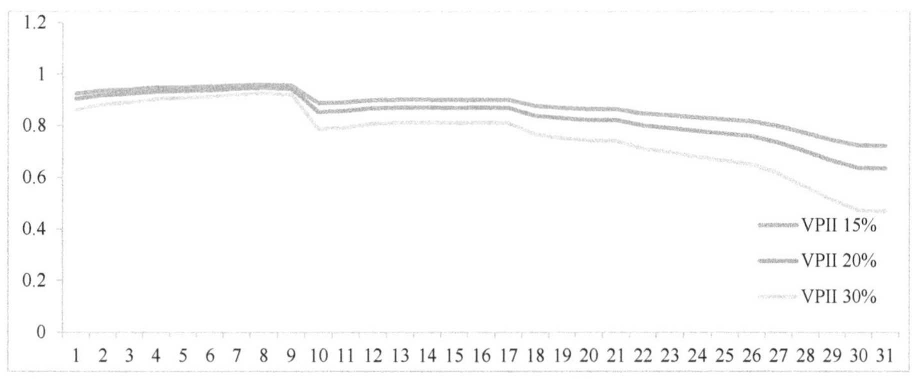

Fig.2.17: Values of voltage profile improvement index at different buses of the system with solar DG penetration at different levels

Table 2. 5: LVDI and WSVPI for different penetration levels.

Bus no	Solar DG at different penetration levels					
	At 15% DG Penetration		At 20% DG Penetration		At 30% DG Penetration	
	LVDI	WSVPI	LVDI	WSVPI	LVDI	WSVPI
1	0	0	0	0	0	0
2	0.001678	0.030292	0.002182	0.025656	0.003117	0.026124
3	0.001713	0.032318	0.002228	0.030549	0.003182	0.031026

4	0.001725	0.037557	0.002247	0.032578	0.003209	0.03306
5	0.001763	0.038934	0.002295	0.037823	0.003282	0.038316
6	0.001778	0.041063	0.002316	0.039203	0.003306	0.039698
7	0.001795	0.045699	0.002336	0.041333	0.003335	0.041833
8	0.001828	0.04907	0.002379	0.045974	0.003402	0.046486
9	0.001861	0.045799	0.00242	0.04935	0.003456	0.049868
10	0.001828	0.043445	0.002379	0.046075	0.003403	0.046586
11	0.00438	0.044706	0.005723	0.044116	0.008243	0.045376
12	0.004401	0.048598	0.005752	0.045381	0.008288	0.046649
13	0.004476	0.049603	0.005843	0.049282	0.008424	0.050572
14	0.004496	0.049633	0.00587	0.05029	0.008461	0.051586
15	0.004496	0.049008	0.005871	0.05032	0.008462	0.051616
16	0.004485	0.049578	0.005858	0.049694	0.008441	0.050985
17	0.004495	0.049033	0.00587	0.050265	0.008461	0.05156
18	0.004485	0.048251	0.005858	0.049719	0.008441	0.051011
19	0.005362	0.051076	0.007001	0.049071	0.010084	0.050612
20	0.006012	0.053005	0.007852	0.051996	0.011311	0.053726
21	0.006501	0.053073	0.008492	0.054001	0.012226	0.055868
22	0.006506	0.058517	0.008493	0.054066	0.012232	0.055936
23	0.008055	0.060414	0.010512	0.059746	0.015131	0.062055
24	0.008697	0.062633	0.011355	0.061742	0.016339	0.064235
25	0.009507	0.064187	0.012409	0.064085	0.017854	0.066807
26	0.010145	0.065669	0.013242	0.065736	0.019046	0.068638
27	0.010788	0.069505	0.014076	0.067313	0.020242	0.070396
28	0.012549	0.07374	0.016368	0.071414	0.023528	0.074994
29	0.014921	0.077607	0.019454	0.076007	0.02794	0.08025
30	0.017365	0.079902	0.022627	0.080239	0.032466	0.085158

| 31 | 0.019208 | 0.079829 | 0.025023 | 0.082737 | 0.03588 | 0.088288 |
| 32 | 0.019325 | 0.045214 | 0.025172 | 0.082826 | 0.036096 | 0.088165 |

Table 2.5. shows the values of Logarithmic Voltage Deviation Index and Weighted Sum of Voltage Profile Index for the penetration levels of 15, 20 and 30 percent respectively. Here the optimal location for solar DG penetration at particular penetration level is chosen based on the maximum value of weighted sum of voltage profile index values obtained from all the buses of the system.

It can be clearly observed in the table that the values of Logarithmic Voltage Deviation Index and Weighted Sum of Voltage Profile Index are less than unity at all busses of the test system and this is evident that the integration of DG has improved the system performance by improving the voltage. Figure 2.17. shows the plot of voltage profile improvement index for solar DG penetration at different levels

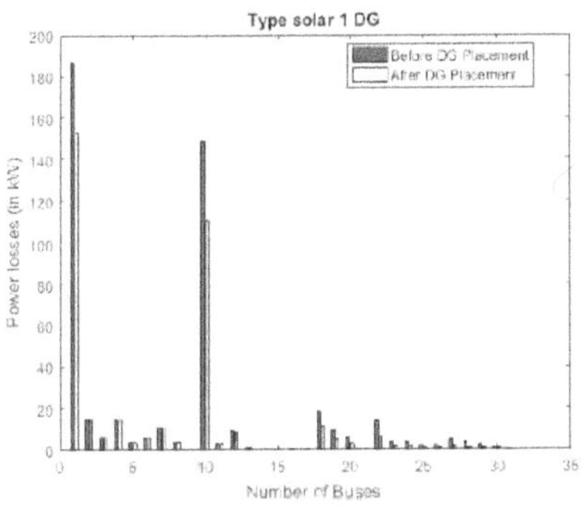

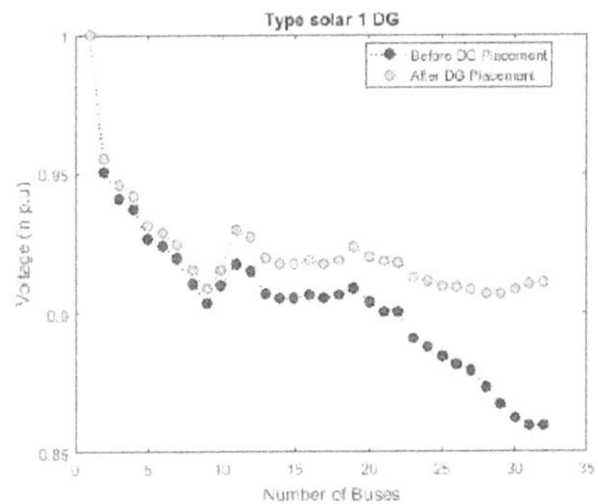

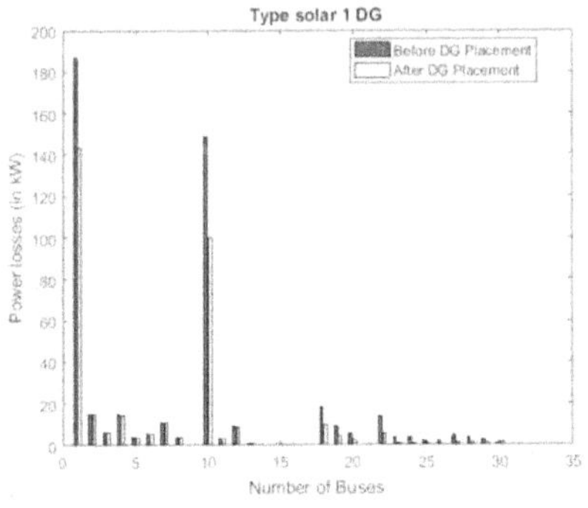

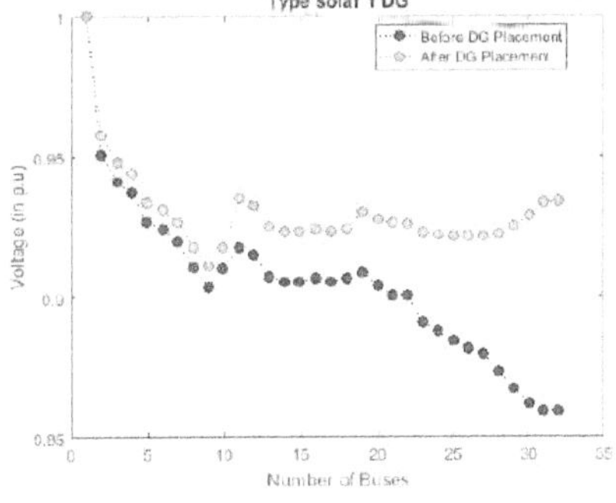

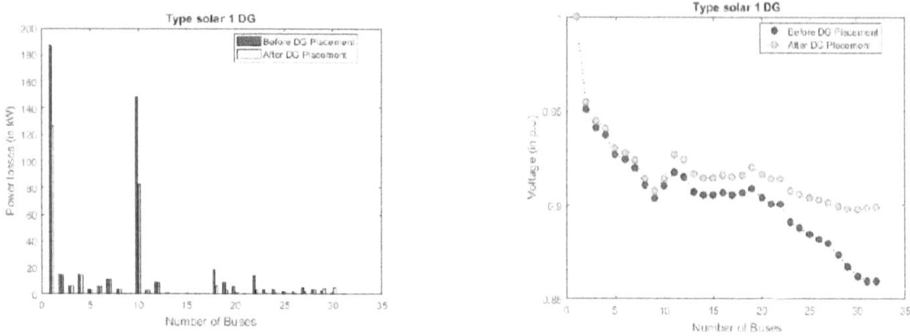

Fig.2.18: Voltage profile and power losses of IEEE33 bus system for integration of solar DG at 15%, 20% and 30% Penetration

The results obtained from the simulation are shown in figure 2.20. the variation of voltage profiles and power losses at different buses of IEEE 33 bus test system for integration of wind powered DG with 15, 20 and 30% penetration.

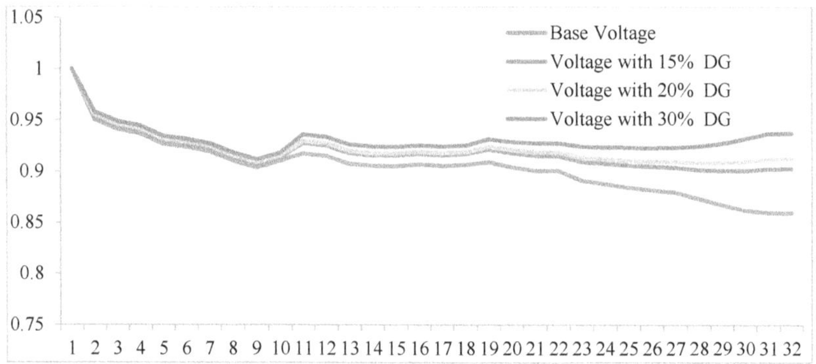

Fig.2.19: Voltage profiles of the system with solar powered DG penetration at different levels

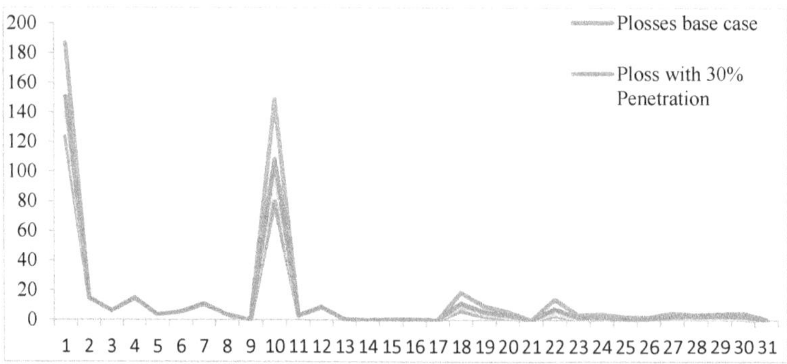

Fig.2.20: Active power loss variation at different buses of the system with solar powered DG penetration at different levels

Fig.2.19 shows the voltage profiles before and after DG placement at different penetration levels say 15, 20 and 30%. The variation of voltage at different buses of the 33-bus test system can be observed in the figure and the voltage profile has been improved with increase in the penetration level. Fig.2.21 shows the system active power losses before and after placement of solar DG at different penetration levels.

Table 2. 6: LVDI and WSVPI for different penetration levels..

Wind DG at different penetration levels

Bus no	At 15% DG Penetration		At 20% DG Penetration		At 30% DG Penetration	
	LVDI	WSVPI	LVDI	WSVPI	LVDI	WSVPI
1	0	0	0	0	0	0
2	0.001464	0.025297	0.002314	0.025722	0.003417	0.026273
3	0.001497	0.030184	0.002361	0.030615	0.003488	0.031179
4	0.001508	0.032209	0.00238	0.032645	0.003522	0.033216
5	0.001544	0.037447	0.002435	0.037892	0.003599	0.038474
6	0.001558	0.038824	0.002456	0.041403	0.003623	0.039857
7	0.00157	0.04095	0.002477	0.046048	0.003658	0.041994
8	0.0016	0.045585	0.002526	0.049422	0.003729	0.046649
9	0.001626	0.048953	0.002563	0.046148	0.003789	0.050035
10	0.0016	0.045685	0.002526	0.044205	0.003729	0.04675
11	0.004745	0.043628	0.0059	0.04547	0.008647	0.045578
12	0.004768	0.044889	0.00593	0.06378	0.008693	0.046852
13	0.004846	0.048783	0.006028	0.089273	0.008837	0.050778
14	0.004871	0.04979	0.006055	0.070382	0.008875	0.051792
15	0.004866	0.049818	0.006055	0.059413	0.008875	0.051823
16	0.00486	0.049195	0.006042	0.049786	0.008854	0.051192
17	0.004866	0.049763	0.006055	0.050357	0.008874	0.051767
18	0.004855	0.049218	0.006038	0.049809	0.00885	0.051215
19	0.005819	0.04848	0.007217	0.049179	0.010569	0.050855
20	0.006538	0.051339	0.008092	0.052116	0.011849	0.053995
21	0.007075	0.053293	0.008747	0.054128	0.012807	0.089158

22	0.007081	0.05336	0.008753	0.054196	0.012813	0.056227
23	0.008786	0.058883	0.010831	0.059905	0.015836	0.062408
24	0.009497	0.060814	0.011693	0.061912	0.017097	0.064613
25	0.010395	0.063078	0.012777	0.064268	0.018674	0.067217
26	0.011101	0.064666	0.013634	0.065932	0.019917	0.069073
27	0.011813	0.066182	0.014491	0.067521	0.021165	0.070857
28	0.013765	0.070112	0.016846	0.071653	0.024582	0.075521
29	0.016403	0.074482	0.020014	0.076287	0.029168	0.080864
30	0.019132	0.080903	0.023272	0.080561	0.03387	0.08586
31	0.021195	0.080823	0.025724	0.083087	0.037408	0.088929
32	0.021326	0.079984	0.025877	0.083178	0.037632	0.089056

Table 2.6 shows the values of Logarithmic Voltage Deviation Index and Weighted Sum of Voltage Profile Index for the penetration levels of 15, 20 and 30 percent respectively. Here the optimal location for wind powered DG penetration at particular penetration level is chosen based on the maximum value of weighted sum of voltage profile index values obtained from all the buses of the system. It can be clearly observed in the table that the values of Logarithmic Voltage Deviation Index and Weighted Sum of Voltage Profile Index are less than unity at all busses of the test system and this is evident that the integration of DG has improved the system performance by improving the voltage.

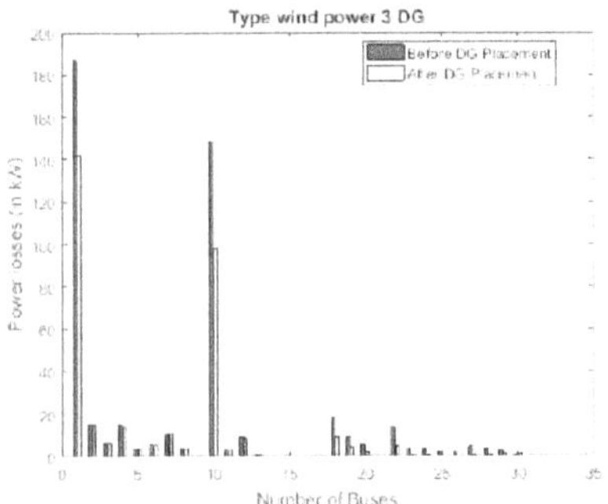

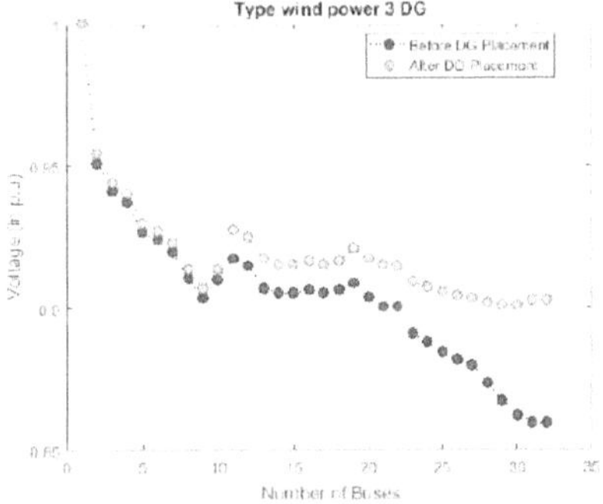

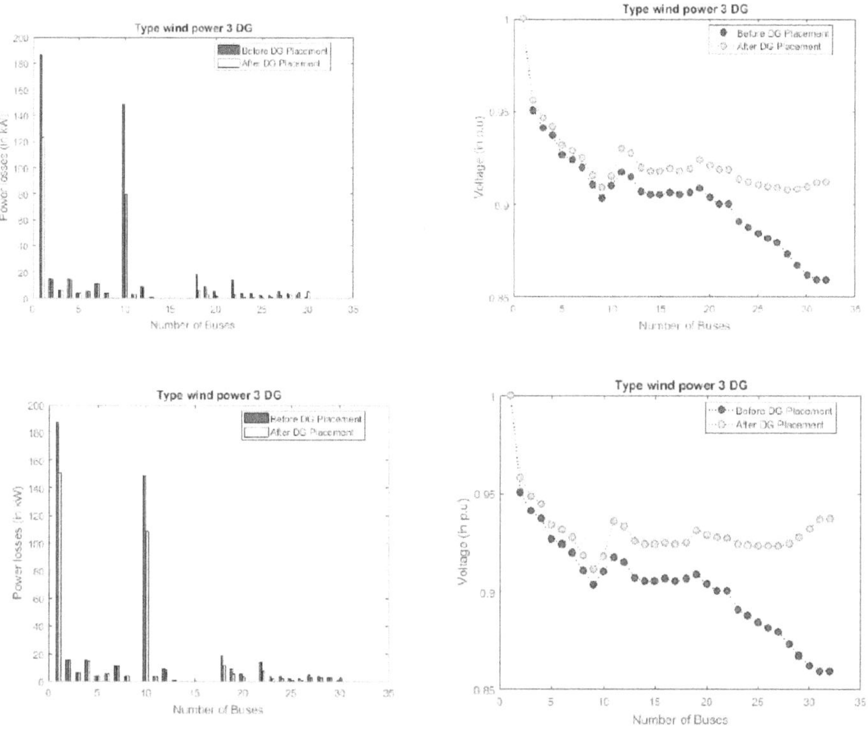

Fig.2.21: Voltage profile and power losses of IEEE33 bus system for integration of wind DG
at 15%, 20% and 30% Penetration

The results obtained from the simulation are shown in figure 2.22. The variation of voltage
profiles and power losses at different buses of IEEE 33bus test system for integration of wind
powered DG with 15, 20 and 30% penetration.

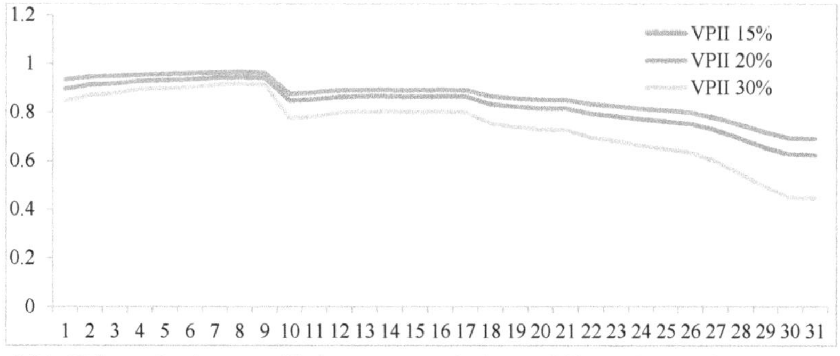

Fig.2.22: Values of voltage profile improvement index at different buses of the system with
wind powered DG penetration at different levels

Figure 2.22 shows the different values of voltage profile improvement index at all the busses of the test system under study. It is evident that the system voltage profile has improved with the integration of DG as the values are less than unity at all the busses.

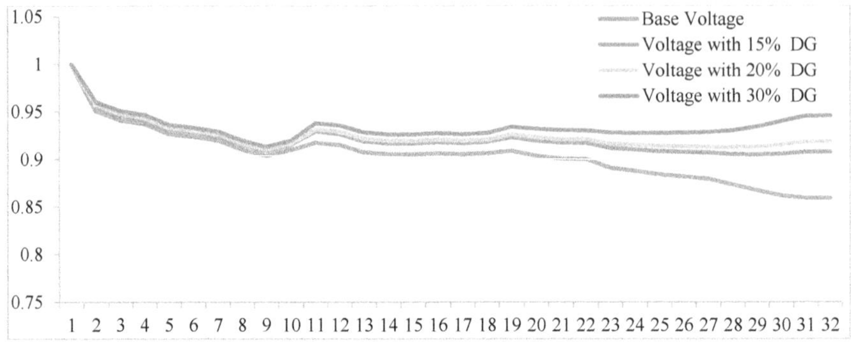

Fig.2.23: Voltage profiles of the system with solar and Wind powered DGs penetration at different levels

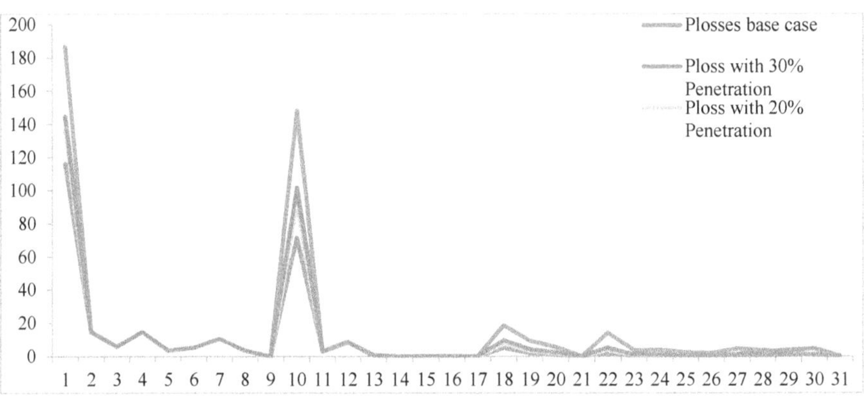

Fig.2.24: Active power loss variation at different buses of the system with wind powered DG penetration at different levels

Fig.2.23 shows the voltage profiles before and after DG placement at different penetration levels say 15, 20 and 30%. The variation of voltage at different buses of the 33-bus test system can be observed in the figure and the voltage profile has been improved with increase in the penetration level. Fig.2.24 shows the system active power losses before and after placement of solar and wind powered DGs at different penetration levels.

Table 2. 7: LVDI and WSVPI for different penetration levels.

Solar and Wind Hybrid DG at different penetration levels

Bus no	At 15% DG Penetration		At 20% DG Penetration		At 30% DG Penetration	
	LVDI	WSVPI	LVDI	WSVPI	LVDI	WSVPI
1	0	0	0	0	0	0
2	0.00225	0.02569	0.002895	0.026013	0.004191	0.02666
3	0.002301	0.030586	0.002957	0.030914	0.00428	0.031575
4	0.00232	0.032615	0.002984	0.032947	0.004316	0.033613
5	0.002374	0.037862	0.00305	0.0382	0.004411	0.038881
6	0.002391	0.03924	0.003073	0.039581	0.004443	0.040266
7	0.002411	0.041371	0.003101	0.041715	0.004482	0.042406
8	0.002459	0.086015	0.003161	0.046365	0.004574	0.047072
9	0.002497	0.049388	0.003213	0.049746	0.004641	0.050461
10	0.00246	0.046115	0.003161	0.046466	0.004575	0.047173
11	0.005447	0.043979	0.006684	0.044597	0.009671	0.04609
12	0.005476	0.045243	0.006721	0.045865	0.009725	0.047367
13	0.005565	0.049142	0.006829	0.049775	0.009882	0.051301
14	0.005591	0.050151	0.006858	0.050784	0.009926	0.052318
15	0.005591	0.050181	0.006859	0.050814	0.009927	0.052349
16	0.005579	0.049555	0.006845	0.050187	0.009904	0.051717
17	0.005591	0.050125	0.006858	0.050759	0.009926	0.052293
18	0.005575	0.049577	0.006841	0.05021	0.0099	0.05174
19	0.006658	0.048899	0.008161	0.04965	0.011803	0.051472
20	0.007465	0.051802	0.009148	0.052644	0.013222	0.054681
21	0.008066	0.053788	0.00988	0.054695	0.014284	0.056897
22	0.008072	0.053856	0.009886	0.054763	0.014286	0.056963
23	0.009979	0.059479	0.012212	0.060596	0.017632	0.063306
24	0.010773	0.061452	0.013181	0.062655	0.019025	0.065578
25	0.011774	0.063767	0.014395	0.065078	0.020762	0.068261
26	0.012558	0.065394	0.015348	0.066789	0.022136	0.070183

27	0.013348	0.066949	0.016312	0.068431	0.02351	0.07203
28	0.01551	0.070985	0.018941	0.0727	0.02727	0.076865
29	0.018419	0.075489	0.022469	0.077515	0.032303	0.082431
30	0.021411	0.079631	0.02609	0.08197	0.037453	0.087651
31	0.023666	0.082058	0.028813	0.084631	0.041317	0.090883
32	0.02372	0.0821	0.028885	0.084682	0.041419	0.09095

Table 2.7 shows the values of Logarithmic Voltage Deviation Index and Weighted Sum of Voltage Profile Index for the penetration levels of 15, 20 and 30 percent respectively. Here the optimal location for solar and wind powered DGs penetration at particular penetration level is chosen based on the maximum value of weighted sum of voltage profile index values obtained from all the buses of the system.

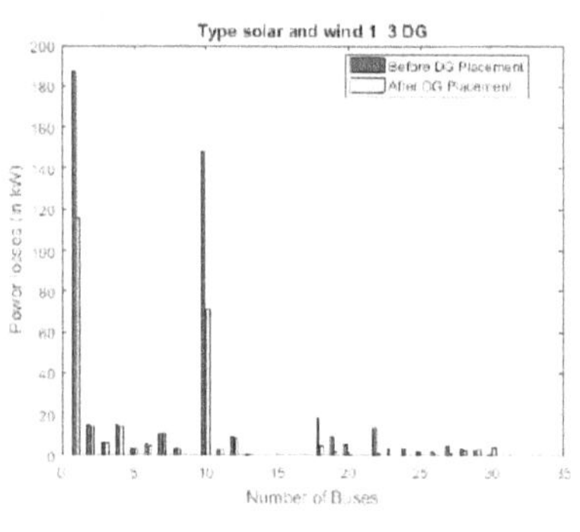

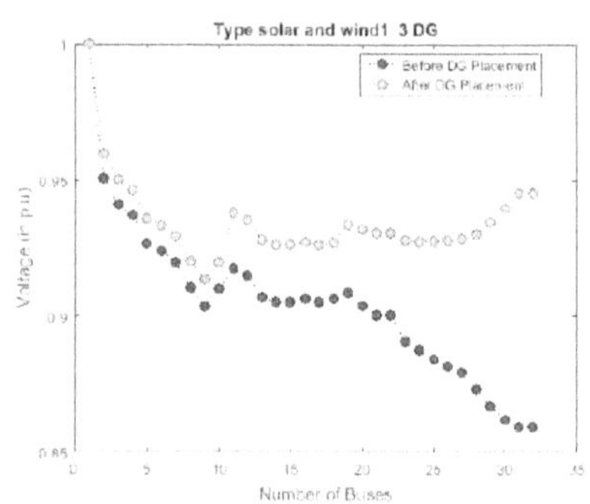

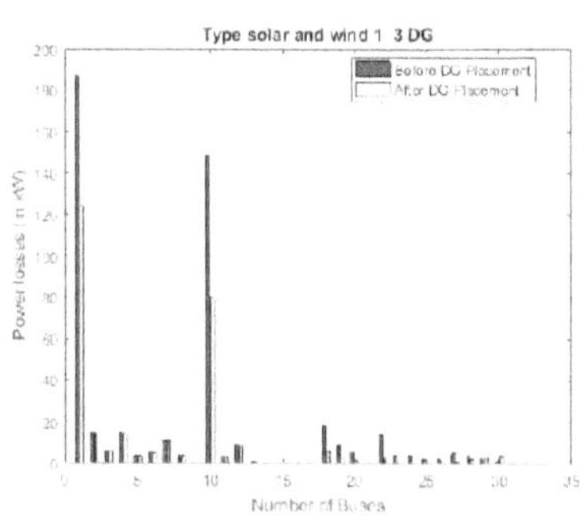

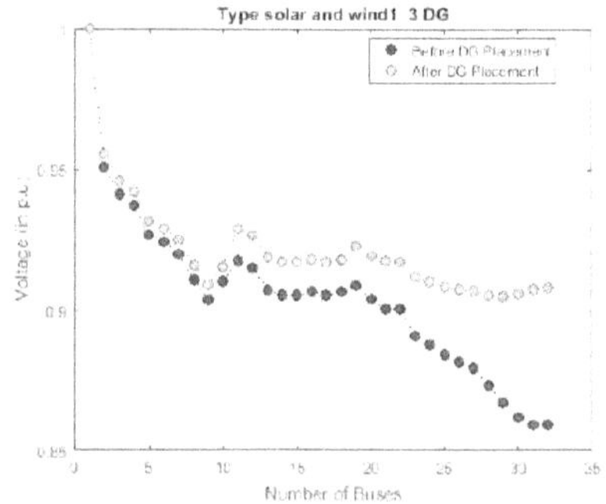

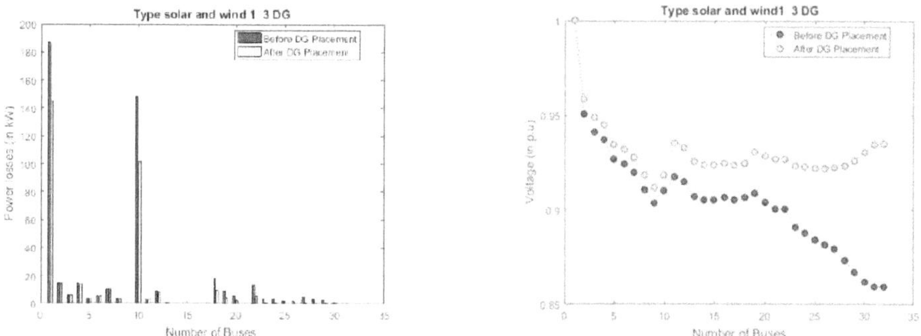

Fig.2.25: Voltage profile and power losses of IEEE33 bus system for integration of solar and wind powered DGs at 15%, 20% and 30% Penetration

The results obtained from the simulation are shown in figure 2.25. The variation of voltage profiles and power losses at different buses of IEEE 33bus test system for integration of solar and wind powered DGs with 15, 20 and 30% penetration.

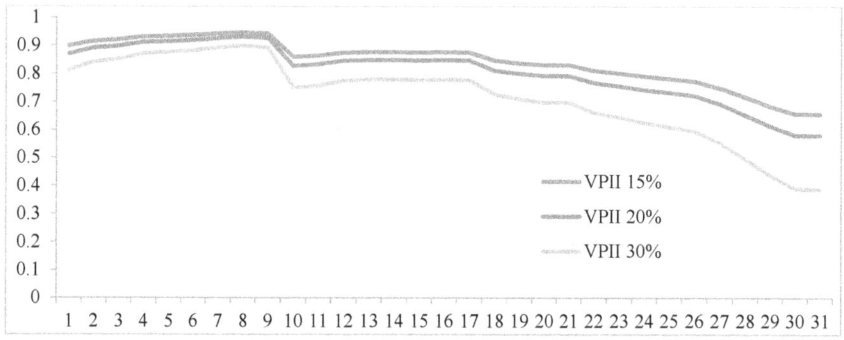

Fig.2.26: Values of voltage profile improvement index at different buses of the system with wind powered DG penetration at different levels

Figure 2.26 shows the different values of voltage profile improvement index at all the busses of the test system under study. It is evident that the system voltage profile has improved with the integration of DG as the values are less than unity at all the busses.

Table 2.8: Summary of voltages and losses for 32 bus practical test system.

DG Penetration Level	Optimal Location	DG type	Active power Losses			Minimum Voltage in p.u	
			Without DG (kW)	With DG (kW)	%PLR	Without DG (p.u)	With DG (p.u)
15 %	30	Solar	477.99	368.56	22.89	0.89529(32)	0.92466

	30	Wind		340.32	28.80		0.93318
	30 21	Solar + wind		321.22	32.79		0.94182
	31	Solar		345.07	27.80		0.94401
20%	13	Wind	477.99	304.3	36.33	0.89529(32)	0.94815
	31 13	Solar + wind		286.84	39.99		0.94989
	31	Solar		314.68	34.165		0.94318
30 %	21	Wind	477.99	281.15	41.18	0.89529(32)	0.94466
	32 08	Solar + wind		260.98	45.40		0.94782

Table 2.8. shows the summary of different DGs penetration levels at 15%, 20% and 30% respectively. Here renewable DGs such as solar, wind and hybrid solar with wind distributed generators are used to improve the system performance by improving the voltage and reducing the losses. The losses are reduced to the maximum value with solar and wind power hybrid combination for the same penetration level.

The losses with the integration of wind powered distributed generation are obtained to be less compared to solar distributed generators, as the solar is capable of generating only active power whereas wind is capable of compensating the reactive power in addition to the real power generation. This research presents the analytical method of optimally allocating the distributed generators using a weighted sum of objective function based on VPI and LVDI methods on the IEEE 33 bus and 32 bus practical radial distribution network. This method utilizes solar and wind types renewable DG's adaptively for finding the best location of the DG's. Results show that the location and penetration levels are the important factors in reducing the losses and improving the voltage profile of the network. The comparison among these types of DG in the IEEE 33 bus and 32 bus practical radial distribution test system showed that DG capable of delivering power gives better system performance. This can be a good guiding step for the operation of renewable DG's

CHAPTER 3

3   To develop an algorithm to find optimal location for different penetration levels of renewable DG's and to analyze the impact of integration on power quality.

3.1 INTRODUCTION

On the basis of modified shuffled frog leap algorithm, a new optimization-based algorithm is utilized and this is done to get renewable DG unit's optimal location. To find the global optimal solution MSFL algorithm is a meta heuristic approach and this is carried out with the help of heuristic search. The information is globally exchanged in the process of evolution by the population and personal memes. The benefits are then aggregated for the localized as well as worldwide searches. On the behavior of group of frogs searching for the location generally forms the basis of MSFLA, these frogs are in search of maximum food availability location. The population of possible solutions is involved in MSFLA by the virtual frogs set. Subsets are made from the set of virtual frogs and these are named as memeplexes. These memeplexes are likely to be perceived as a parallel culture frog sets which are trying to attain the similar goals.

An individual frog improves the frog leaping and, in this way, to attain the goal there are more chances. Each frog holds different ideas in all the memeplexes and one frog's idea is utilized to infect the other frog's idea. Local search or memetic evolution step is the name given to the process of data transferring between the memeplex frogs. Memetic evolution steps when once defined, there is a shuffling of the frogs and then these frogs are reorganized so that there is an enhancement of the memeplex quality. The memetic quality is improved by the shuffling once the infection is caused and this ensures the evolution of a culture in direction of a specific interest. Until a needed convergence is reached the process of evolution and shuffling are repeated.

The modification with respect to the placement of different types of DG is done by selecting the condition of load (normal and heavy) on the system and availability of type of DG. Behavior of local and global search is used for the identification of load condition and the virtual frogs for identifying the best location for the selected type of DG with selected percentage of penetration as these will identify the maximum availability of food. With above modification power losses, number of sag and swell buses are reduced, and possible to keep the voltage stability index value as minimum as possible.

3.2 Literature review

The SFLA improvement for the solution of problems of multiple objective optimizations, explorations either local or global, local optima trap avoiding, reducing the time needed for computation, initial population quality improvement etc are all included. On the statistical outcomes the measured enhancements in SFLA are based and in this the modification which researchers have been doing in the past. In the end, the SFLA is addressed by quantitative validations as a robust algorithm which in various applications is employed and other optimization algorithms are outperformed by it. [101]. Since there are many algorithms, for global optimal solutions SFLA is suggested via various heuristic functions [103]. It is regarded as one of the efficient ways in solving the problems of multi objectives as well as complex organizations. In the section to follow, the SFLA with basic concepts and framework along with its upcoming generations are discussed so that it becomes stronger.

The process of shuffling as well as localized searches takes place until the criteria of stopping is fulfilled. [104]. The individual memetic quality is enhanced by the memetic evolution. The memetic approach is transferred in the local search in case of SFLA in the process of shuffling and in this way amongst the global the memetic is transferred. [105]. Better performance is there for the SFLA in the large size generation for the global search and memeplexes and higher time is needed for their computation. Along with this, the process of shuffling also increases the memetic quality on the basis of different cultures of memeplexes. Between local and global research when the information is exchanged, the flexibility and robustness of the outcomes is ensured. [102]. In the other form alternative to it, is the determinacy ways and the integration of SFLA [105][106]. The messages are exchanged in determinacy approach successfully in the algorithm and also it ensures the robustness and algorithmic flexibility [105].

To SFLA there are some other alternative studies for the solution of combinatorial as well as problems of multi objective optimizations with objective imprecise functions while to find optimum solution basic SFLA has some limitations. In this way, to improve the basic SFLA performance the research workers continue to work. Either on the algorithm the enhancement is done or this is done by the SFLA modification or sometimes by the SLFA or HSFLA hybridization. With the other evolutionary algorithms, the HSFLA and SFLA are integrated. As a result of it, SFLA in a continuous pattern makes it difficult to resolve the problems of optimization since they are large scaled and complicated [109]. Some previously suggested structures for the modified SFLA and MSFLA are used by the research workers to handle these kinds of optimization problems and to improve the overall functioning. One more step in considered by the research workers into the MSFLA suggested [110]. With the help of this step, a number of frogs for each memeplexes with higher level of fitness are selected to make a selected memeplex performing well independently at the local exploration. The rate of convergence and the time of processing are increased by the sub memeplexes.

In SFLA to overcome the premature convergence, [108] and [107] suggested a modern form of equation and they did it by applying search acceleration as a new factor. As a constant value the factor is positive as well as time's nonlinear function, exploration can be balanced by it either globally or locally keeping in mind the fundamentals of exploration and search aspects in the feasible solution regions [111] which to solve the direct SFLA discrete problems is modified. Hence the strategy used is the threshold selection and in frog leaping operation it is employed in solving the problems such as multi user detection. Problems such as pre mature convergence are defeated along with other problems such as lower speed of convergence, for the improvement of MSFLA accuracy optimization [101] and two important modifications are employed making use of SFLA. The internal learning effect is benefited making use of a new division strategy of memeplexes to make a uniform performance at the memeplex. This also cancels the sub memeplex construction to allow better chances of learning to frogs and this is the way to increase global exploration. The basic SFLA has some limitations and to overcome these limitations in the phase of evolution [109] recommended to make use of best information at local level for the frog leaping and to pass them on between the individual frogs. As a result of this [112] applied MSFLA for the solutions of problem related to multi objectives. The algorithmic technique which was offered is considered to handle the solutions of non-dominated kinds. [113] along with this applied another approach which was suggested by [114. 115] for the proper handling of multi objective concerns. As per the outcomes of computation, the MSFLA presented is highly efficient and competitive to solve the complicated concerns as compared to the active programming.

The basis SFLA is modified for the rate of convergence enhancement and to solve the concerns regarding multi objectives [118]. In MSFLA presented here, to hold the non-dominated solutions the external repository is defined. To manage the repository size the fuzzy clustering method is also applied. It has been shown in the outcomes of evaluations that as compared to PSO and GA the MSFLA is more effective. Along with this new mutation was applied by [116] and this was to reduce the time of computation and to raise the quality of solutions, this also avoid the local optima trapping. A modern MSFLA was used [118] and this was done by the modification of the SFLA frog leaping and Tribe MSFLA was employed as well so that prematurity issues can be avoided. In Tribe MSFLA each memeplex is considered as a tribe.

3.3 Methodology

Step 1: solve the feeder line flow by reading the line and load data of the system making use of a method known as load flow. Branch current load flow method is used in this work.

Step 2: the best DG locations should be determined for the buses making use of a method known as index vector method.

Step 3: for each individual P fitness is supposed to be calculated by initializing the population or solutions and by the generation of solutions P for the random population.

Step 4: P population sorting for the fitness divide P in the descending order for each memeplex for each m memeplex. The best and worst frogs are determined.

Step 5: DG locations and population solutions are randomly generated with the help of

$$sol\ (i,:) = LVDI_{min} + (LVDI_{max} + LVDI_{min}) * rand\ (1,d)$$

Step 6: for general population the active power loss is determined by the low flow performance

Step 7: for the voltage deviation the DG location is selected as a best solution by the improvement of position of worst frog.

Step 8: on the basis of p using equation generation of new and local population solutions.

Step 9: by the performance of load flow determination of the voltage profiles for the population updated.

Step 10: the current proper solutions can be replaced with the values updated if there are lesser obtained losses and voltage profile as compared to the current optimal solution, in other case switch to step number seven.

Step 11: print the outcomes if you reach the maximum iterations.

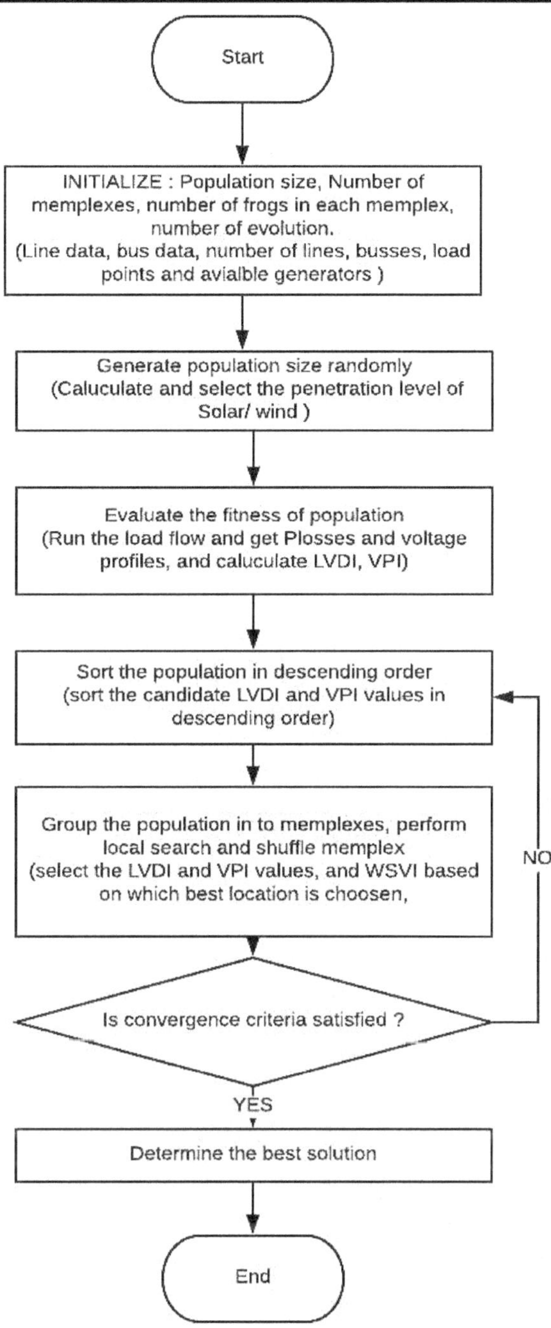

Fig 3.1: Flowchart of the algorithm for MSFLA

3.4 Problem Formulation

In this work, the DGs are employed to enhance the system performance in power systems. The optimal placing and fixed penetration of renewable DGs for "power quality improvement" using MSFLAP is proposed. The main purpose of DGs location is to improve the voltage profile by reducing the deviation index and to minimize the "real power loss" in the system, so as to improve the system subjected to various operational and security constraints.

3.4.1 The "real power loss" can be expressed by the relation

$$P = \sum_{ij=1}^{N}(P_{ij} + P_{ji}) \dots\dots\dots\dots\dots\dots\dots\dots\dots\dots\dots\dots\dots\dots\dots\dots\dots\dots3.1$$

Where, Pij and Pji are the real power flows from bus i to j and from bus j to i respectively.

Power quality can be improved by optimally allocating the available distributed generation sources using the Modified shuffled frog leap algorithm. The inclusion of DG in an appropriate way will give a better profile of the voltage and reduce the system losses, hence improving the system performance by reducing the power quality problems.

$$P_{Gj}^{min} \leq P_{Gj} \leq P_{Gj}^{max} \dots\dots\dots\dots j = 1, 2, \dots . Ng$$

$$Q_{Gj}^{min} \leq Q_{Gj} \leq Q_{Gj}^{max} \dots\dots\dots\dots j = 1, 2, \dots . Ng$$

$$P_{DG}^{min} \leq P_{DG} \leq P_{DG}^{max}$$

$$V_{n}^{min} \leq V_{n} \leq V_{n}^{max} \dots\dots\dots\dots n = 1, 2, \dots . Nd$$

The DG's can be allocated optimally by minimizing the objective function given in equation meeting all the constraints limitations. A metaheuristic algorithm is proposed in this chapter.

3.4.2 Power Quality assessment Indices

The system performance will be depending on the quality of the power supplied and in various cases it is specified by using System Average RMS Frequency Index (SARFI) [3], it presents the average number of RMS variations over the assessment period to the customer served, and it calculates the number of busses effected by voltage sag by crossing the lower limit of 0.9.p.u. If the limit is fixed to 90% irrespective of the duration it is represented as SARFI-90.

SARFI is an acronym for the System Average RMS Variation Frequency Index. It is a power quality index that provides a count or rate of voltage sags, swells, and/or interruptions for a system. The SARFI index was first described by Brooks in [3]. The size of the system is scalable: it can be defined for a single monitor, a single customer service, a feeder, one or more substations, or an entire power delivery system.

SARFI-X corresponds to a count or rate of voltage sags, interruptions and/or swells below/above a specified voltage threshold. For example, SARFI-70 considers voltage sags and interruptions that are below 70% of the reference voltage. SARFI-110 considers voltage swell and interruptions that are above 110% of the reference voltage. SARFI indices are meant to assess short-duration rms variation events only, meaning that only those events are included in its computation with durations less than the minimum duration of a sustained interruption as defined by IEEE Std 1159, which is one minute [3].

$$SARFI = \frac{\textit{Total number of buses experiencing voltage sag (Nsag)}}{\textit{Total number of customers served}} \dots\dots\dots\dots\dots\dots\dots\dots 3.2$$

And the improvement index of the same is represented as

$$SARFII = \frac{SARFI_{withDG}}{SARFI_{withoutDG}} \dots\dots\dots\dots\dots\dots\dots\dots\dots\dots\dots\dots\dots\dots\dots\dots\dots 3.3$$

Where,

SARFI = System Average RMS Frequency Index

SARFII = System Average RMS Frequency Improvement Index

SARFI $_{without\,DG}$= System Average RMS Frequency Improvement Index without DG

SARFI $_{with\,DG}$ System Average RMS Frequency Improvement Index with DG

Nsag = Total number of buses experiencing voltage sag.

$$N_{sag} = \sum_{i=1}^{N_{bus}} \begin{cases} 1 & if \quad 0.1pu < V_i < 0.9\,p.u \\ 0 & otherwise \end{cases} \dots\dots\dots\dots\dots\dots\dots\dots 3.4$$

Where, N_{sag} corresponds to busses which fall below the lower limit under the heavy loaded condition. N_{bus} corresponds to number of buses.

$$N_{Swell} = \sum_{i=1}^{N_{bus}} \begin{cases} 1 & if \quad V_i > 1.04\,p.u \\ 0 & otherwise \end{cases} \dots\dots\dots\dots\dots\dots\dots\dots 3.5$$

Where, N_{Swell} corresponds to busses which fall below the lower limit under the heavy loaded condition. N_{bus} corresponds to number of buses.

$$VSI = \frac{Total\ number\ of\ Buses\ experienceing\ voltage\ sag}{Total\ number\ of\ b\ uses\ in\ the\ network} \dots\dots\dots\dots\dots\dots\dots\dots\dots\dots\dots\dots\dots\dots\dots\dots 3.6$$

VSI= Voltage Sag Index

3.4.3. The objective functions are as fallows

F_1= Min (P)

F_2= Min (N_{sag} and N_{swell}) there by satisfying condition of F3

F_3 = Min (VSI)

3.5. Simulation Results and discussions

This section shows the results obtained after the execution of MSFLA on IEEE 33 bus, 32 bus practical radial bus of Mysore, Karnataka, India using MATLAB simulation. For the network considered the following three cases are analyzed and for each case, three scenarios are considered.

Case-1 : Allocation of only Solar DG units

 Case-2 : Allocation of only wind powered DG units

Case-3 : Allocation of both solar and wind powered DG units

Scenario -1: With penetration level of 15%

Scenario -2: With penetration level of 30%

Scenario -3: With penetration level of 50%

The system is loaded with additional 60% of its load at all busses with the consideration of constant power model. This heavy loaded system is used for the analysis to examine the power quality indices using case4, 5 and case6. These cases are evaluated with integration of single DG unit. Under this condition, the system losses will increase to 559.66 kW compared with 202.8 kW of 100% load. Number of buses which falls below the lower limit of 90% regardless of sag duration are treated as sag buses. Then SARFI is calculated from total number of sag buses and also SARFII is calculated to show the power quality improvement after DG placement.

Case -4: Optimal Allocation of only solar DG unit under heavy load condition

Case -5: Optimal Allocation of only wind powered DG unit under heavy load condition

Case -6: Optimal Allocation of both solar and wind powered DG unit under heavy load condition

Scenario -1: DG Integration at an optimal location

Scenario -2: DG integration at two optimal locations

Case 6 is not analyzed with first scenario as two DGs are integrated with their optimal locations hence, scenario 2 is used for the analysis

3.4.3   CASE 1: IEEE 33 BUS TEST SYSTEM

This method is applied to the standard IEEE 33 bus system operated at 12.66kV level of voltage and consists of 33 busses, 32 branches, with 3715kW and 2300kVar of power respectively as shown in Fig.3.2. The basic data of the network that is the line data and the branch data are given in the appendix A. The base case load flow is obtained where the power losses without the connection of the DG's are 202.68kW and 143.22kVAr respectively.

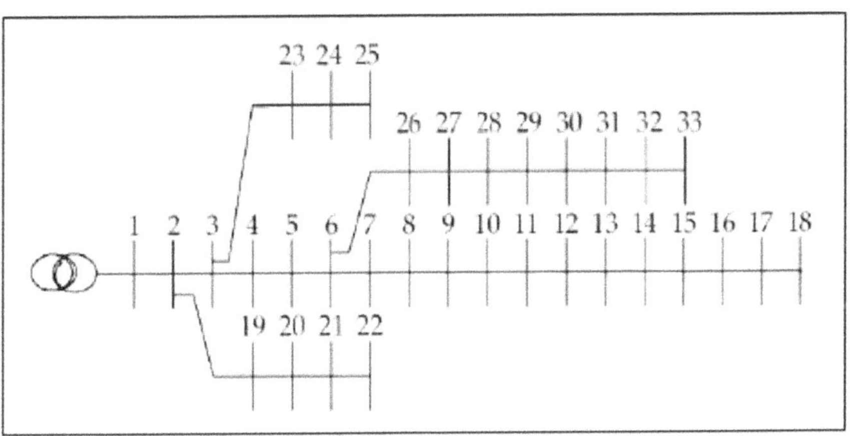

Fig.3.2: Single diagram of IEEE 33 bus test system

Particulars	Abbreviations
1: Active power losses in kW (Base case)	APL (kW)
2: Active Power losses with DG in kW	APL W DG (kW)
3: Net Power loss reduction in kW	NPLR (kW)
4: Net power loss reduction in %	NPLR (%)
5: Minimum voltage without DG in pu.	Min V w/o DG (pu.)
6: Minimum voltage with DG in pu.	Min V WDG (pu.)

Case-1, Scenario-1: 15% penetration of solar DG

The results obtained from the developed MSFLA for the placement of solar DG unit are shown in Table 3.1. The total losses of the system losses reduced to 155.73KW from 202.68KW with a net loss reduction of 46.95KW and percentage losses are reduced by 23.16%.

Table 3.1: Penetration of solar DG for scenario 1

Particulars	MSFLA Solar DG
1: Best location for the scenario 1	Bus 6
2: 15% Penetration in kW	557.5
3: APL (kW)	202.68
4: APL W DG (kW)	155.73
5: NPLR (kW)	94.6
6: NPLR (%)	23.16
7: Min V w/o DG (p.u.)	Bus 18
	0.91309
8: Min V WDG (p.u.)	Bus 18
	0.92178

Case-1, Scenario-2: 30% penetration of solar DG

The results obtained from the developed MSFLA for the placement of solar DG unit are shown in Table 3.2. The total losses of the system losses reduced to 123.71KW from 202.68KW with a net loss reduction of 123.71KW and percentage losses are reduced by 38.96%.

Table 3.2: Penetration of solar DG for scenario 2

Particulars	MSFLA Solar DG
1: Best location for the scenario 2	Bus 6
2: 30 % Penetration in kW	1114.5
3: APL (kW)	202.68
4: APL with SDG (kW)	123.71
5: NPLR (kW)	78.97
6: NPLR (%)	38.96
7: Min V w/o DG (p.u.)	Bus 18 0.91309
8: Min V with SDG (p.u.)	Bus 18 0.93145

Case-1, Scenario-3: 50% penetration of solar DG

The results obtained from the developed MSFLA for the placement of solar DG unit are shown in Table 3.3. The total losses of the system losses reduced to 113.4KW from 202.68KW with a net loss reduction of 89.28KW and percentage losses are reduced by 44.05%.

Table 3.3: Penetration of solar DG for scenario 3

Particulars	MSFLA Solar DG
1: Best location for the scenario 3	Bus 27
2: 50% Penetration in kW	1857
3: APL (kW)	202.68
4: APL W DG (kW)	113.4

5: NPLR (kW)	89.28
6: NPLR (%)	44.05
7: Min V w/o DG (p.u.)	Bus 18
	0.91309
8: Min V with DG (p.u.)	Bus 18
	0.94083

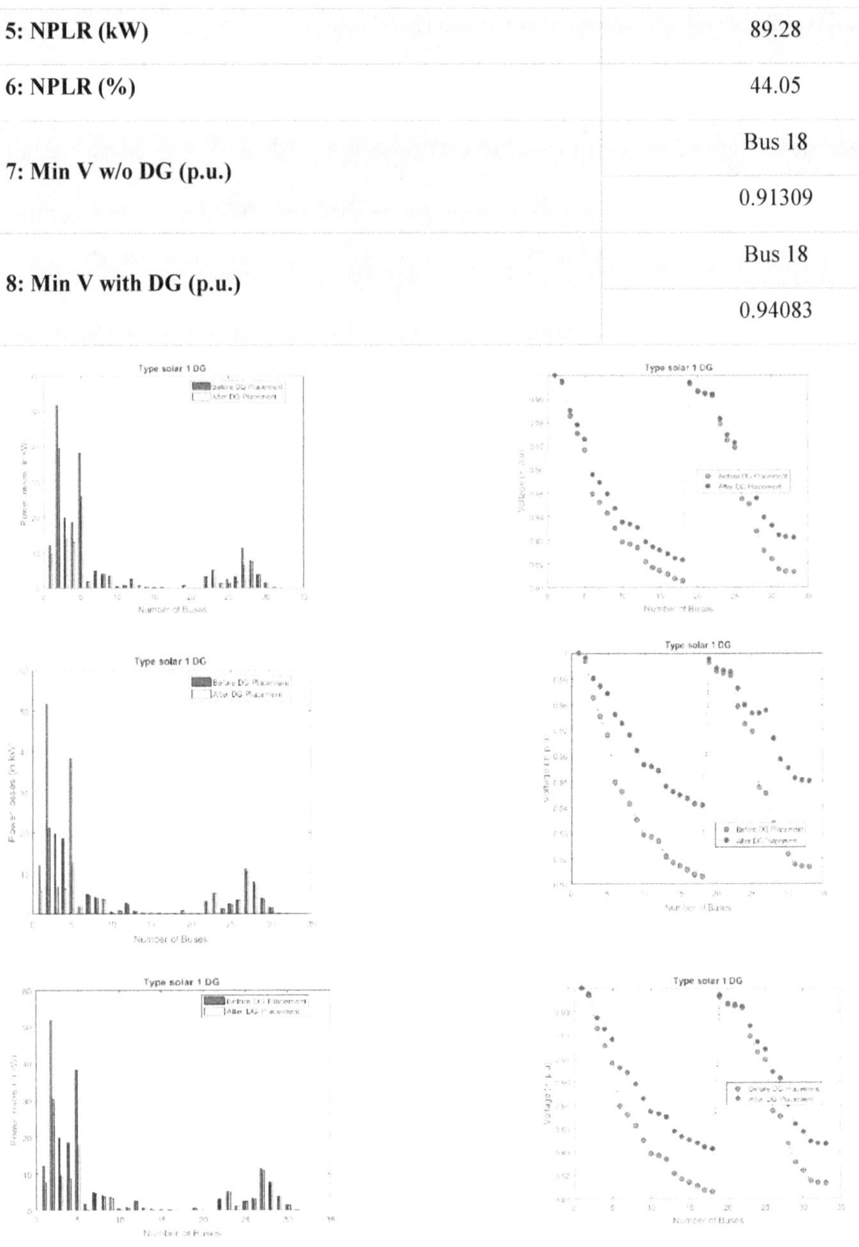

Fig. 3.3: Power losses and voltage profiles of the IEEE33 bus system with 15, 30 and 50% of solar DG Penetration at an optimal location using MSFLA.

From the Fig.3.3, it can be observed that the losses are drastically reduced with the integration of solar DG as the penetration level increases. The losses are reduced the maximum with the 50% penetration level.

Case-2, Scenario-1: 15% penetration of wind powered DG : The results obtained from the developed MSFLA for the placement of wind powered DG unit are shown in Table 3.4. The total losses of the system losses reduced to 142.84KW from 202.68KW with a net loss reduction of 59.84KW and percentage losses are reduced by 29.52%.

Table 3.4: Penetration of wind DG for scenario 1

Particulars	MSFLA Wind powered DG
1: Best location for the scenario 1	Bus 7
2: 15% Penetration in kW and kVAR	557kW, - 345kVAR
3: APL (kW)	202.68
4: APL W DG (kW)	142.84
5: NPLR (kW)	59.84
6: NPLR (%)	29.52
7: Min V w/o DG (p.u.)	Bus 18 0.91309
8: Min V W DG (p.u.)	Bus 18 0.94526

Case-2, Scenario-2: 30% penetration of wind powered DG

The results obtained from the developed MSFLA for the placement of wind powered DG unit are shown in Table 3.5. The total losses of the system losses reduced to 119.58KW from 202.68KW with a net loss reduction of 83.09KW and percentage losses are reduced by 41.39%.

Table 3.5: Penetration of wind powered DG for scenario 2

Particulars	MSFLA Wind powered DG
1: Best location for the scenario 2	Bus 33
2: 30% Penetration in kW and kVAR	1.1145kW, -690kVAR

3: APL (kW)	202.68
4: APL W DG (kW)	119.58
5: NPLR (kW)	83.09
6: NPLR (%)	41.39
7: Min V w/o DG (p.u.)	Bus 18 0.91309
8: Min V WDG (p.u.)	Bus 18 0.92691

Case-2, Scenario-3: 50% penetration of wind powered DG

The results obtained from the developed MSFLA for the placement of wind powered DG unit are shown in Table 3.6. The total losses of the system losses reduced to 108.08KW from 202.68KW with a net loss reduction of 94.60KW and percentage losses are reduced by 46.67%.

Table 3.6: Penetration of wind powered DG for scenario 3

Particulars	MSFLA Wind powered DG
1: Best location for the scenario 3	Bus 12
2: 50% Penetration in kW and kVAR	1.857 kW, -1150kVAR
3: APL (kW)	202.68
4: APL W DG (kW)	108.08
5: NPLR (kW)	94.6
6: NPLR (%)	46.67
7: Min V w/o DG (p.u.)	Bus 18 0.91309
8: Min V WDG (p.u.)	Bus 33 0.94668

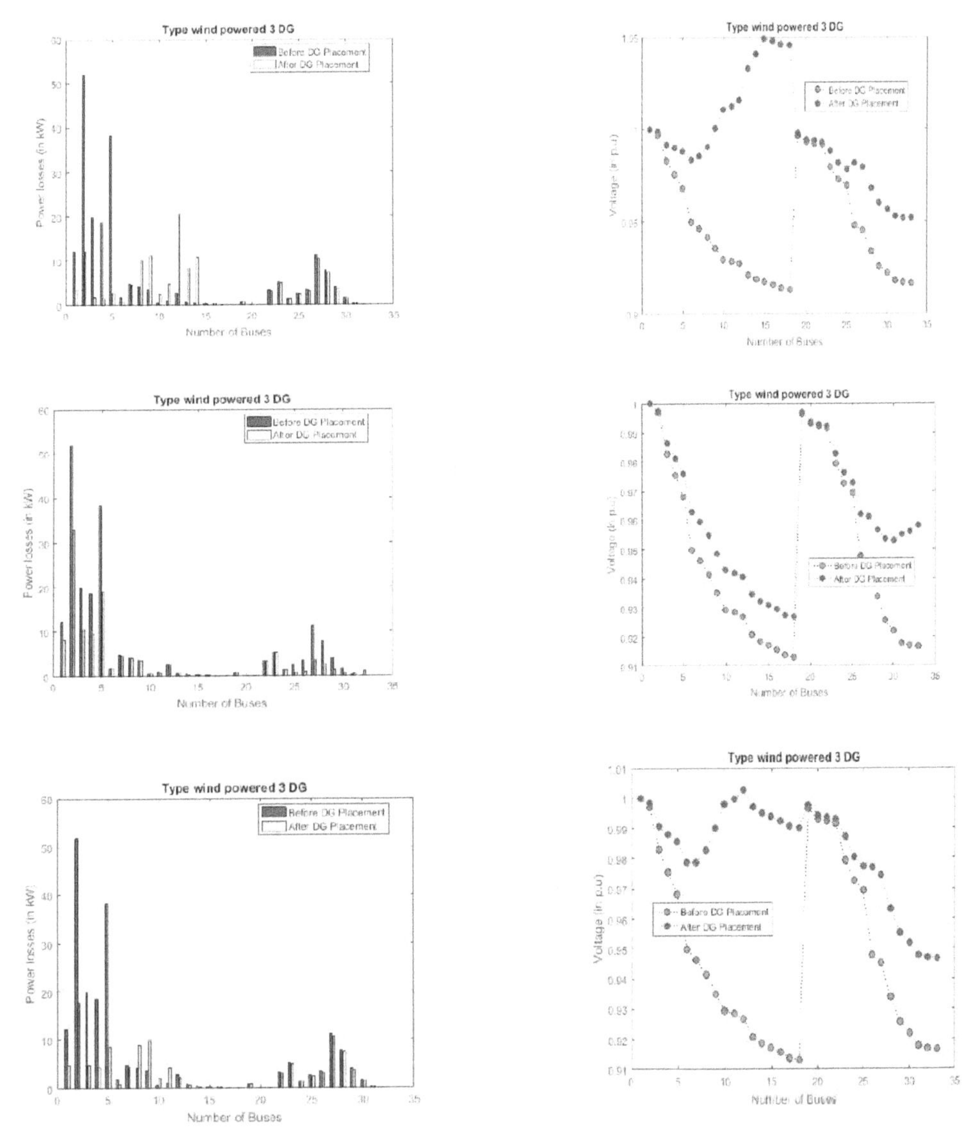

Fig. 3.4: Power losses and voltage profiles of the IEEE33 bus system with 15, 30 and 50 % of wind powered DG Penetration at an optimal location using MSFLA.

Fig.3.4 shows the active power loss and voltage profiles before and after integration of wind powered DGs with different penetration levels. In comparison with solar DGs better voltage profile and reduction of losses are obtained when integrated with wind powered DGs for the same penetration levels.

Case-3, Scenario-1: 15% penetration of solar and wind powered hybrid DGs

The results obtained from the developed MSFLA for the placement of wind powered DG unit are shown in Table 3.7. The total losses of the system losses reduced to 134.78KW from 202.68KW with a net loss reduction of 67.9KW and percentage losses are reduced by 33.96%. Compared to first two cases here the losses are reduced to the maximum with the same penetration level.

Table 3.7: Penetration of solar and wind powered hybrid DG for scenario 1

Particulars	MSFLA Solar and Wind Powered hybrid DGs
1: Best locations for scenario 1	Bus 2 - solar Bus30 - wind
2: APL (kW)	202.68
3: APL W DG (kW)	134.78
4: NPLR (kW)	67.9
5: NPLR (%)	33.96
6: Min V w/o DG (p.u.)	Bus 18 0.91309
7: Min V WDG (p.u.)	Bus 18 0.93741

Case-3, Scenario-2: 30% penetration of solar and wind powered hybrid DGs

The results obtained from the developed MSFLA for the placement of wind powered DG unit are shown in Table 3.8. The total losses of the system losses reduced to 105.94KW from 202.68KW with a net loss reduction of 96.74KW and percentage losses are reduced by 47.72%. Compared to first two cases here the losses are reduced to the maximum with the same penetration level.

Table 3.8: Penetration of solar and wind powered hybrid DG for scenario 2

Particulars	MSFLA Solar and Wind Powered hybrid DGs
1: Best locations for scenario 2	Bus 30 - wind Bus02 - solar
2: APL (kW)	202.68

3: APL W DG (kW)	105.94
4: NPLR (kW)	96.74
5: NPLR (%)	47.72
6: Min V w/o DG (p.u.)	Bus 18 0.91309
7: Min V WDG (p.u.)	Bus 18 0.94188

Case-3, Scenario-3: 50% penetration of solar and wind powered hybrid DGs

The results obtained from the developed MSFLA for the placement of wind powered DG unit are shown in Table 3.9. The total losses of the system losses reduced to 89.39KW from 202.68KW with a net loss reduction of 113.29KW and percentage losses are reduced by 55.90%. Compared to first two cases here the losses are reduced to the maximum with the same penetration level.

Table 3.9: Penetration of solar and wind powered hybrid DG for scenario 3

Particulars	MSFLA Solar and Wind Powered hybrid DGs
1: Best locations for scenario 3	Bus 30 - wind Bus02 - solar,
2: APL (kW)	202.68
3: APL W DG (kW)	89.39
4: NPLR (kW)	113.29
5: NPLR (%)	55.90
6: Min V w/o DG (p.u.)	Bus 18 0.91309
7: Min V WDG (p.u.)	Bus 18 0.9569

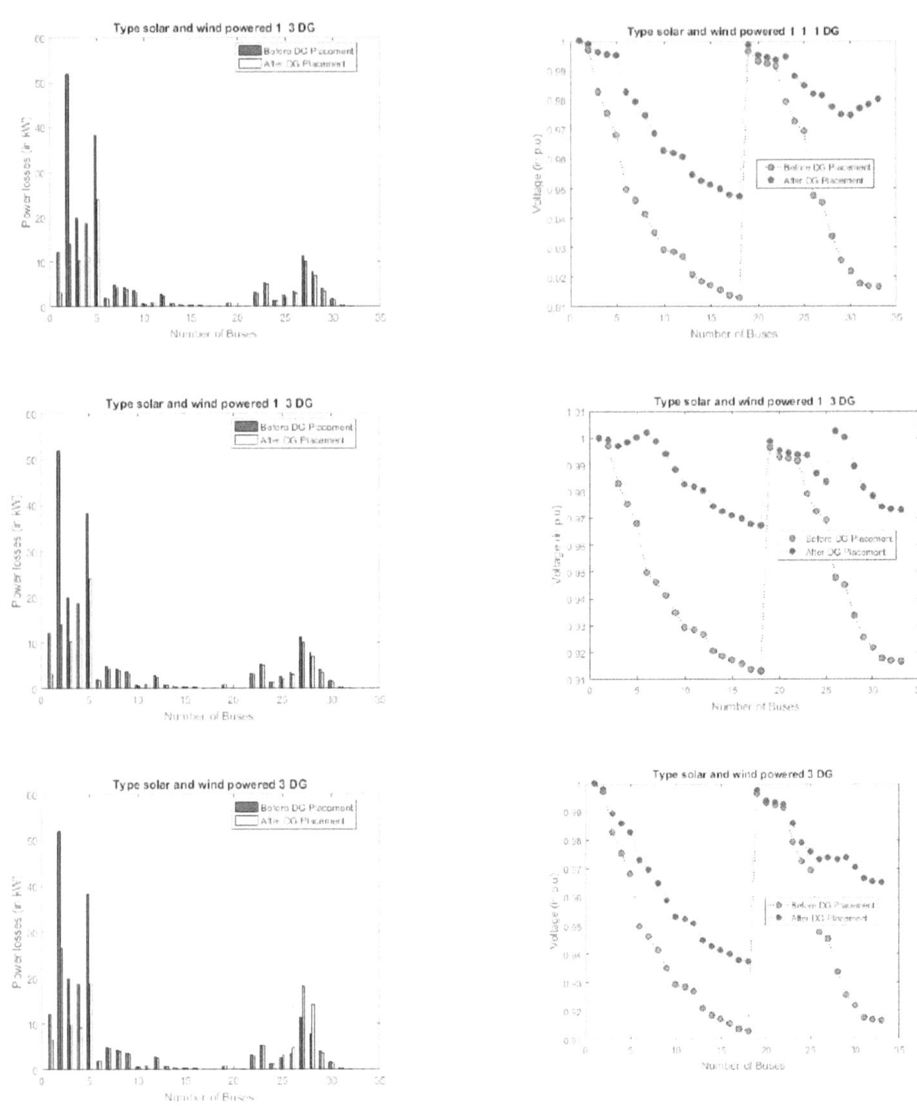

Fig. 3.5: Power losses and voltage profiles of the IEEE33 bus system with 15, 30 and 50 % of solar and wind powered hybrid DG Penetration at an optimal location using MSFLA.

Fig.3.5 shows the active power loss and voltage profiles before and after integration of solar and wind powered hybrid DGs with different penetration levels. In comparison with solar DGs and wind powered DGs better voltage profile and reduction of losses are obtained when integrated with hybrid DGs for the same penetration levels.

Table.3.10 shows the comparison of different methods with their obtained results for IEEE 33 bus RDS system with proposed MSFLA method. Here MSFLA method is compared with the penetration level of 50% with the available methods with the optimal penetration of the literature. The results of the different methods are taken from the literature [71, 72, 102].

Proposed FPA results are compared with other methods were total losses with integration hybrid DG at 50% penetration level obtained to be 89.39kW i.e., 55.90% reduction of losses which provides much better operation of the systems when compared to other techniques.

Table 3.10: Comparison table for allocation of DG using different techniques

Method	P loss with DG (kW)	Loss reduction in %	Minimum voltage in p.u (Bus)	(bus) DG location	Optimal DG size (MW)
GA[71]	107.1	49.71	0.9810 (25)	11	1.5
				29	0.4228
				30	1.0714
PSO [71]	105.35	50.06	0.9806(30)	13	0.9816
				32	0.8297
				8	1.1768
GA/PSO [71]	103.4	50.99	0.9808(25)	32	1.2
				16	0.863
				11	0.925
MSFLA	102.8	50.72	0.97526	07	2.3156

Case-4, Scenario-1: optimal placement of a Solar DG unit

Table 3.11 shows the results obtained from developed MSFLA algorithm for allocation of a Solar DG unit for Heavy load condition. Voltage sag index before (VSI) and SARFI before installation of DG is obtained to be 0.4848 and 0.0040858 and after DG integration it got reduced to 0.3030 and 0.0038304, also Voltage sag improvement index and SARFII obtained to be 0.3030and 0.068750. The system losses reduced to 482 kW from 559.66KW with a net loss reduction of 77.6 kW and system losses got reduced by 13.87%.

Table 3.11: optimal allocation of solar DG for scenario1 under heavy load

Solar (Single DG Penetration)	Heavy load
1: Optimal Node (location)	Bus 18
2 : Optimal size (kW)	330
3: APL w/o DG (kW)	559.66
4: APL W DG (kW)	482
5: APLR (kW)	77.6
6: APLR (%)	13.87
7: Min V w/o DG (p.u.)	Bus 18 0.85493
8: Min V W DG (p.u)	Bus 18 0.86684
Voltage sag index before DG placement	0.4848
Number of sag buses (Nsag), Before DG placement	16
Number of sag buses (Nsag), After DG placement	10
Voltage sag index after DG placement	0.3030
SARFI,Before DG placement	0.0040858
SARFI,after DG placement	0.0038304
SARFII	0.9374908

Case-4, Scenario-2: optimal placement of two solar DG units

Table 3.12 shows the results obtained from developed MSFLA algorithm for allocation of two Solar DG units for Heavy load condition. SARFI before installation of DG is obtained to be 0.0040858 and after DG integration it got reduced to 0.0024723, also SARFII obtained to be 0.068750.

The system losses reduced to 432.47 kW from 559.66kW with a net loss reduction of 127.19 kW and system losses got reduced by 22.72 %. In the base case with heavy load condition the value of voltage sag index obtained to be 0.4848 and after integration of DG at the optimal location it reduced to 0.2424 and Voltage sag improvement index obtained to be 0.2424

Table 3.12: optimal allocation of solar DG for scenario 2 under heavy load

Solar (Two DGs Penetration)	Heavy load
1: Optimal Node (location)	Bus 29 Bus 12
2 : Optimal size in kW	309.3 337.44
3: APL w/o DG (kW)	559.66
4: APL W DG (kW)	432.47
5: APLR (kW)	127.19
6: APLR (%)	22.72
7: Min V w/o DG (p.u.)	Bus 18 0.85493
8: Min V WDG (p.u)	Bus 18 0.87236
Voltage sag index before DG placement	0.4848
Number of sag buses (Nsag), Before DG placement	16
Number of sag buses (Nsag), After DG placement	8
Voltage sag index after DG placement	0.2424
SARFI,Before DG placement	0.0040858
SARFI,after DG placement	0.0024723
SARFII	0.605095

Case-5, Scenario-1: optimal placement of a wind Powered DG unit

Table 3.13 shows the results obtained from developed MSFLA algorithm for allocation of a Solar DG unit for Heavy load condition. Voltage sag index before (VSI) and SARFI before installation of DG is obtained to be 0.4848 and 0.0040858 and after DG integration it got reduced to 0.2424 and 0.0033123, also Voltage sag improvement index and SARFII obtained to be 0.2424 and 0.8106857. The system losses reduced to 441.01 kW from 559.66KW with a net loss reduction of 118.64 kW and system losses got reduced by 21.20%.

Table 3.13: optimal allocation of Wind powered DG for scenario1 under heavy load

Wind Power (Single DG Penetration)	Heavy load
1: Optimal Node (location)	Bus 31
2 : Optimal size	536 kW 618.66 kVAR
3: APL w/o DG (kW)	559.66
4: APL W DG (kW)	441.01
5: APLR (kW)	118.64
6: APLR (%)	21.20
7: Min V w/o DG (p.u.)	Bus 18 0.85493
8: Min V W DG (p.u)	Bus 18 0.87014
Voltage sag index before DG placement	0.4848
Number of sag buses (Nsag), Before DG placement	16
Number of sag buses (Nsag), After DG placement	8
Voltage sag index after DG placement	0.2424
SARFI,Before DG placement	0.0040858
SARFI,after DG placement	0.0033123
SARFII	0.8106857

Case-2, Scenario-2: optimal placement of two wind powered DG units

Table 3.14 shows the results obtained from developed MSFLA algorithm for allocation of two Solar DG units for Heavy load condition. SARFI before installation of DG is obtained to be 0.0040858 and after DG integration it got reduced to 0.0021237, also SARFII obtained to be 0.5197758.

The system losses reduced to 381.28 KW from 559.66KW with a net loss reduction of 178.38 KW and system losses got reduced by 31.87 %. In the base case with heavy load condition the value of voltage sag index obtained to be 0.4848 and after integration of DG at the optimal location it reduced to 0.1515 and Voltage sag improvement index is obtained to be 0.1515.

Table 3.14: optimal allocation of Wind powered DGs for scenario 2 under heavy load

Wind power (Two DGs Penetration)	Heavy load
1: Optimal Node (location)	Bus 29 Bus 12
2 : Optimal size	321.21 kW, 389.3 kVAR 444.51 kW,494.51 kVAR
3: APL w/o DG (kW)	559.66
4: APL W DG (kW)	381.28
5: APLR (kW)	178.38
6: APLR (%)	31.87
7: Min V w/o DG (p.u.)	Bus 18 0.85493
8: Min V WDG (p.u)	Bus 18 0.88101
Voltage sag index before DG placement	0.4848
Number of sag buses (Nsag), Before DG placement	16
Number of sag buses (Nsag), After DG placement	5
Voltage sag index after DG placement	0.1515
SARFI,Before DG placement	0.0040858

| SARFI,after DG placement | 0.0021237 |
| SARFII | 0.5197758 |

Case-6, Scenario-1: optimal placement of solar and wind powered DG unit

Table 3.15 shows the results obtained from developed MSFLA algorithm for allocation of a Solar DG unit for Heavy load condition. Voltage sag index before (VSI) and SARFI before installation of DG is obtained to be 0.4848 and 0.0040858 and after DG integration it got reduced to 0.1212 and 0.0027321, also Voltage sag improvement index and SARFII obtained to be 0.1212 and 0.6785231. The system losses reduced to 318.33 KW from 559.66KW with a net loss reduction of 241.32 KW and system losses got reduced by 43.12%.

Table 3.15: optimal allocation of solar and Wind powered DG under heavy load

Solar and Wind Power (DG Penetration)	Heavy load
1: Optimal Node (location)	Bus 06 Bus 31
2 : Optimal size	309.21 kW 512.21 kW ,602.12 kVAR
3: APL w/o DG (kW)	559.66
4: APL W DG (kW)	318.33
5: APLR (kW)	241.32
6: APLR (%)	43.12
7: Min V w/o DG (p.u.)	Bus 18 0.85493
8: Min V WDG (p.u)	Bus 18 0.88203
Voltage sag index before DG placement	0.4848
Number of sag buses (Nsag), Before DG placement	16
Number of sag buses (Nsag), After DG placement	4
Voltage sag index after DG placement	0.1212

SARFI,Before DG placement	0.0040858
SARFI,after DG placement	0.0027231
SARFII	0.6785231

3.4.4 Case Study 2: 32 bus practical radial distribution network

The proposed method is tested on the 32 bus radial distribution network (Bhuvaneshwari feeder, mysuru, India) a practical system. The loads are assumed to be constant with the voltage level of 11kv and 20 MVA Base. 3715kW and 2300kVar is the total active and reactive power of the system respectively. Fig 3.7 shows the diagram of the network considered. The basic data of the network that is the line data and the branch data are given in appendix A. The nodes with the highest amount of deviation are identified and tabulated in the table. The base case load flow is obtained where the power losses without the penetration of the renewable DG's are obtained to be 1.1109kW.

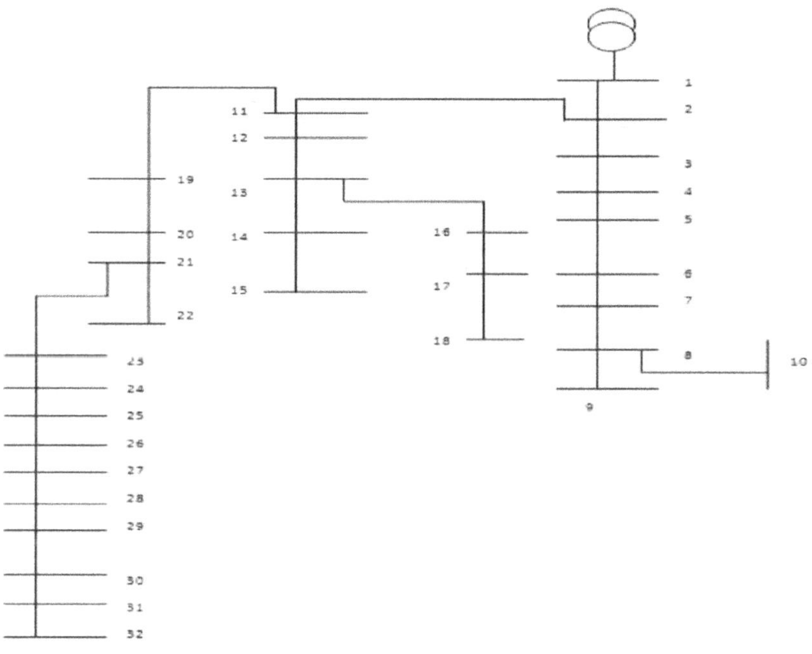

Fig 3.7: Single line diagram of 32 bus practical test system.

Case-1: Scenario-1: 15% penetration of solar DG

The results obtained from the developed MSFLA for the placement of solar DG unit are shown in Table 3.16. The total losses of the system losses reduced to 324.58 kW from 477.99 kW with a net loss reduction of 94.6 kW and percentage losses are reduced by 32.09%.

Table 3.16: Penetration of solar DG for scenario 1

Particulars	MSFLA Solar DG
1: Best location for the scenario 1	Bus 32
2: 15% Penetration in kW	0.5575
3: APL (kW)	477.99
4: APL W DG (kW)	324.58
5: NPLR (kW)	94.6
6: NPLR (%)	32.09
7: Min V w/o DG (p.u.)	Bus 32 0.85952
8: Min V W DG (p.u.)	Bus 09 0.91605

Case-1: Scenario-2: 30% penetration of solar DG

The results obtained from the developed MSFLA for the placement of solar DG unit are shown in Table 3.17. The total losses of the system losses reduced to 300.51 KW from 477.99KW with a net loss reduction of 177.47 KW and percentage losses are reduced by 37.31%.

Table 3.17: Penetration of solar DG for scenario 2

Particulars	MSFLA Solar DG
1: Best location for the scenario 2	Bus 32
2: 30 % Penetration in kW	1.1145
3: APL (kW)	477.99

4: APL W DG (kW)	300.51
5: NPLR (kW)	177.47
6: NPLR (%)	37.13
7: Min V w/o DG (p.u.)	Bus 32 0.85952
8: Min V W DG (p.u.)	Bus 09 0.91342

Case-1: Scenario-3: 50% penetration of solar DG

The results obtained from the developed MSFLA for the placement of solar DG unit are shown in Table 3.18. The total losses of the system losses reduced to 209.78 KW from 477.99 KW with a net loss reduction of 268.20 KW and percentage losses are reduced by 43.39%.

Table 3.18: Penetration of solar DG for scenario 3

Particulars	MSFLA Solar DG
1: Best location for the scenario 3	Bus 27
2: 50% Penetration in kW	1.857
3: APL (kW)	477.99
4: APL W DG (kW)	209.78
5: NPLR (kW)	268.20
6: NPLR (%)	43.89
7: Min V w/o DG (p.u,)	Bus 32 0.85952
8: Min V W DG (p.u.)	Bus 09 0.92577

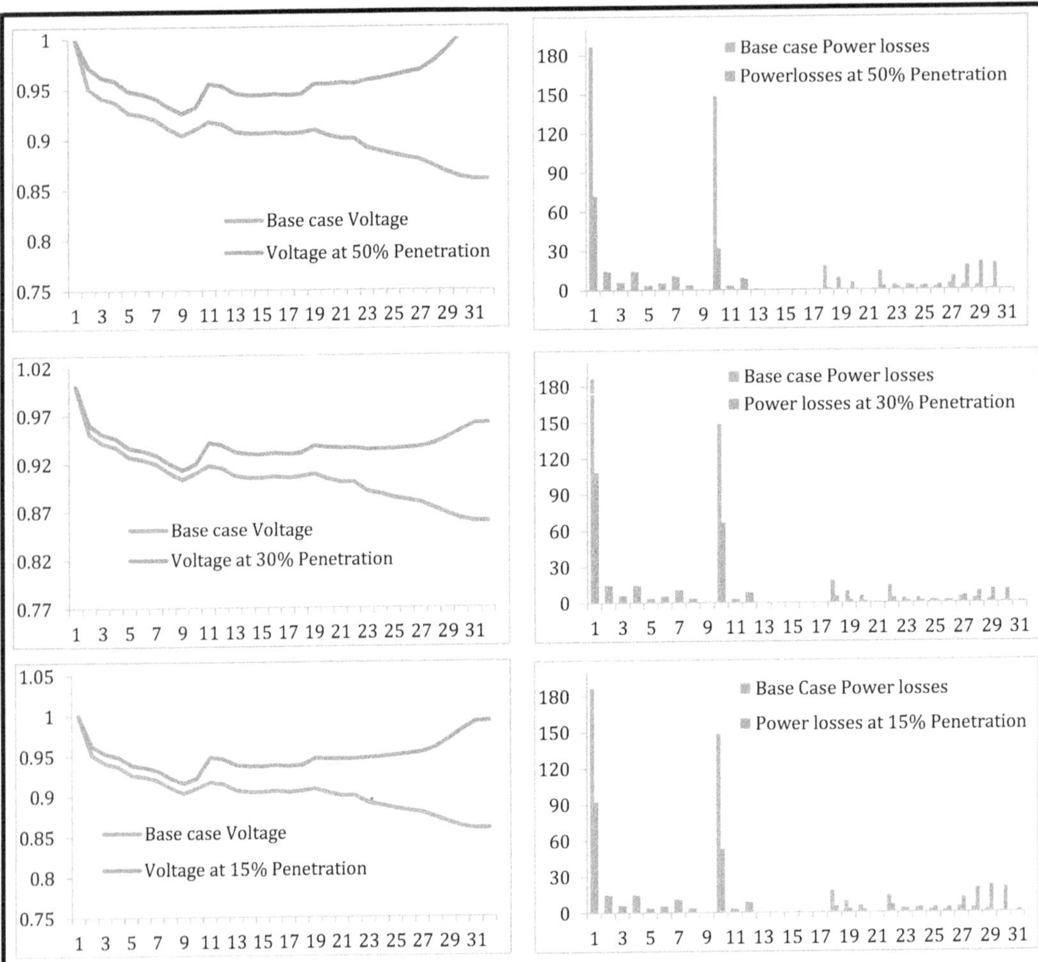

Fig. 3.8: Power losses and voltage profiles of the 32-bus practical system with 50, 30 and 15 % of solar DG Penetration at an optimal location using MSFLA.

From the Fig.3.8, it can be observed that the losses are drastically reduced with the integration of solar DG as the penetration level increases. The losses are reduced the maximum with the 50% penetration level.

Case-2, Scenario-1: 15% penetration of wind powered DG

The results obtained from the developed MSFLA for the placement of wind powered DG unit are shown in Table 3.19. The total losses of the system losses reduced to 324.58 kW from 477.99 kW with a net loss reduction of 94.6 kW and percentage losses are reduced by 32.09%.

Table 3.19: Penetration of wind DG for scenario 1

Particulars	MSFLA Wind powered DG
1: Best location for the scenario 1	Bus 32
2:APL (kW)	477.99
3: APL W DG (kW)	324.58
4: NPLR (kW)	94.6
5: NPLR (%)	32.09
6: Min V w/o DG (p.u.)	Bus 32 0.85952
7: Min V W DG (p.u.)	Bus 09 0.91605

Case-2, Scenario-2: 30% penetration of wind powered DG

The results obtained from the developed MSFLA for the placement of wind powered DG unit are shown in Table 3.20. The total losses of the system losses reduced to 261.98 kW from 477.99 KW with a net loss reduction of 216.01 kW and percentage losses are reduced by 45.19%.

Table 3.20: Penetration of wind powered DG for scenario 2

Particulars	MSFLA Wind powered DG
1: Best location for the scenario 2	Bus 31
2: APL (kW)	477.99
3: APL W DG (kW)	261.98
4: NPLR (kW)	216.01
5: NPLR (%)	45.19
6: Min V w/o DG (p.u.)	Bus 32 0.85952
7: Min V W DG (p.u.)	Bus 09

0.92822

Case-2, Scenario-3: 50% penetration of wind powered DG

The results obtained from the developed MSFLA for the placement of wind powered DG unit are shown in Table 3.21. The total losses of the system losses reduced to 238.11 kW from 477.99 kW with a net loss reduction of 239.88kW and percentage losses are reduced by 50.18%.

Table 3.21: Penetration of wind powered DG for scenario 3

Particulars	MSFLA Wind powered DG
1: Best location for the scenario 3	Bus 31
2: APL (kW)	477.99
3: APL W DG (kW)	238.11
4: NPLR (kW)	239.88
5: NPLR (%)	50.18
6: Min V w/o DG (p.u.)	Bus 32 0.85952
7: Min V W DG (p.u.)	Bus 09 0.92277

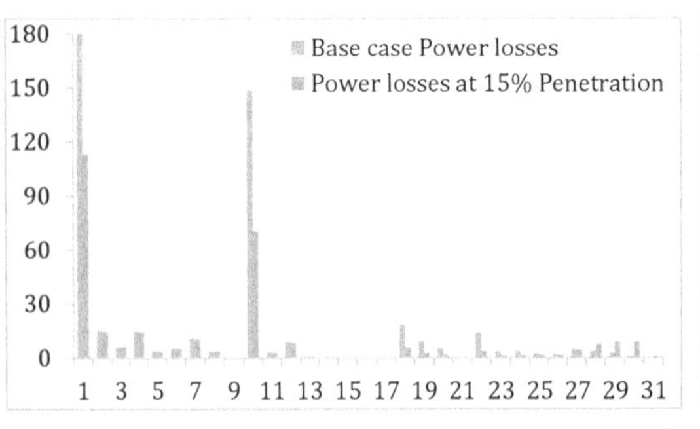

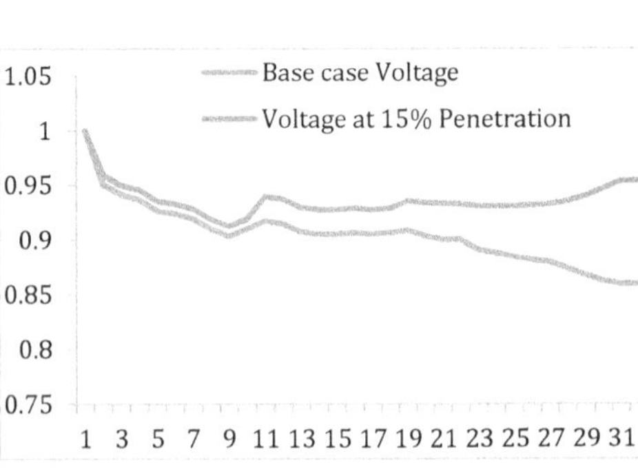

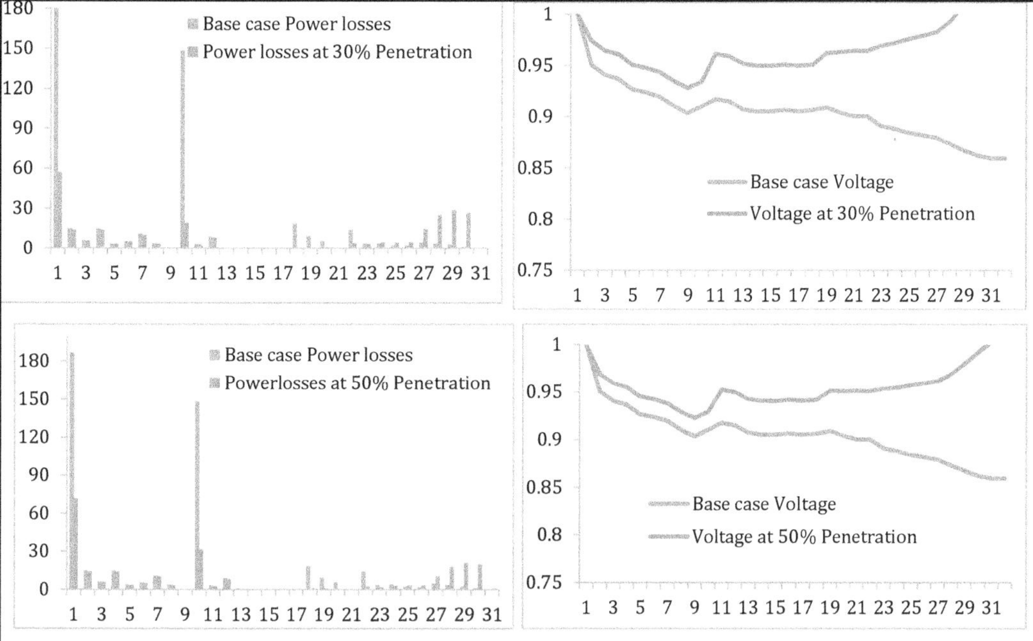

Fig. 3.9: Power losses and voltage profiles of the 32-bus practical system with 15, 30 and 50 % of wind powered DG Penetration at an optimal location using MSFLA.

Fig.3.9 shows the active power loss and voltage profiles before and after integration of wind powered DGs with different penetration levels. In comparison with solar DGs better voltage profile and reduction of losses are obtained when integrated with wind powered DGs for the same penetration levels.

Case-3, Scenario-1: 15% penetration of solar and wind powered hybrid DGs

The results obtained from the developed MSFLA for the placement of wind powered DG unit are shown in Table 3.22. The total losses of the system losses reduced to 303.6 KW from 477.99 kW with a net loss reduction of 174.39 kW and percentage losses are reduced by 36.48%. Compared to first two cases here the losses are reduced to the maximum with the same penetration level.

Table 3.22: Penetration of solar and wind powered hybrid DG for scenario 1

Particulars	MSFLA Solar and Wind Powered hybrid DGs
1: Best locations for scenario 1	Bus 2 - Solar Bus30 - Wind

Particulars	
2: APL (kW)	477.99
3: APL W DG (kW)	303.6
4: NPLR (kW)	67.9
5: NPLR (%)	36.48
6: Min V w/o DG (p.u.)	Bus 32 0.85952
7: Min V W DG (p.u.)	Bus 9 0.91304

Case-3, Scenario-2: 30% penetration of solar and wind powered hybrid DGs:

The results obtained from the developed MSFLA for the placement of wind powered DG unit are shown in Table 3.23. The total losses of the system losses reduced to 256.84 kW from 477.99 kW with a net loss reduction of 221.15 kW and percentage losses are reduced by 46.26%.

Table 3.23: Penetration of solar and wind powered hybrid DG for scenario 2

Particulars	MSFLA Solar and Wind Powered hybrid DGs
1: Best locations for scenario 2	Bus 2 - Solar Bus30 - Wind
2: APL (kW)	477.99
3: APL W DG (kW)	256.84
4: NPLR (kW)	221.15
5: NPLR (%)	46.26
6: Min V w/o DG (p.u.)	Bus 32 0.85952
7: Min V W DG (p.u.)	Bus 9 0.92114

Case-3, Scenario-3: 50% penetration of solar and wind powered hybrid DGs

The results obtained from the developed MSFLA for the placement of wind powered DG unit are shown in Table 3.24. The total losses of the system losses reduced to 212.94 KW from 477.99 KW with a net loss reduction of 265.04 KW and percentage losses are reduced by 55.90%.

Table 3.24: Penetration of solar and wind powered hybrid DG for scenario 3

Particulars	MSFLA Solar and Wind Powered hybrid DGs
1: Best locations for scenario 3	Bus 2 - Solar Bus30 - Wind
2: APL (kW)	477.99
3: APL W DG (kW)	212.94
4: NPLR (kW)	265.04
5: NPLR (%)	54.55
6: Min V w/o DG (p.u.)	Bus 32 0.85952
7: Min V W DG (p u.)	Bus 9 (0.92305 pu)

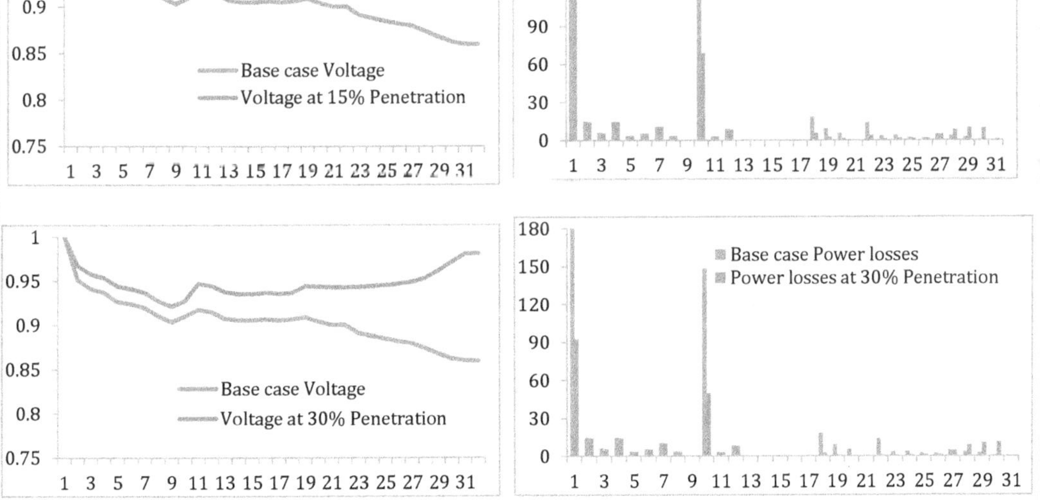

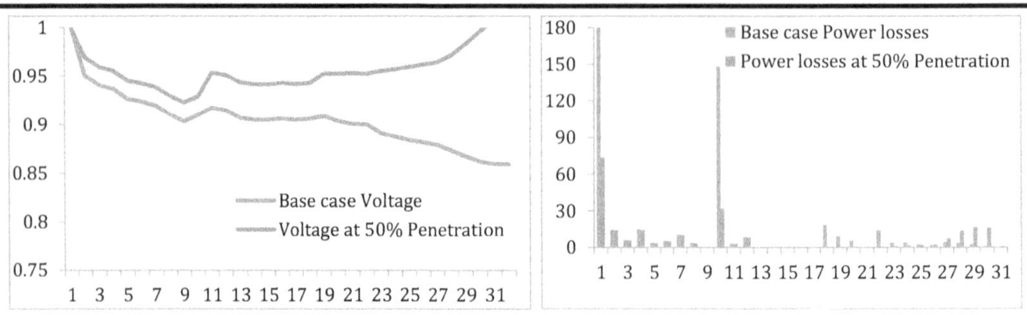

Fig. 3.10 Power losses and voltage profiles of the 32-bus practical system with 15, 30 and 50 % of solar and wind powered hybrid DG Penetration at an optimal location using MSFLA.

Fig.3.10 shows the active power loss and voltage profiles before and after integration of solar and wind powered hybrid DGs with different penetration levels. In comparison with solar DGs and wind powered DGs better voltage profile and reduction of losses are obtained when integrated with hybrid DGs for the same penetration levels.

Case-4, Scenario-1: optimal placement of a Solar DG unit

Table 3.25 shows the results obtained from developed MSFLA algorithm for allocation of a Solar DG unit for Heavy load condition. VSI and SARFI before installation of DG is obtained to be 0.9375 and 0.0086207 and after DG integration it got reduced to 0.71875 and 0.0078966, also Voltage sag improvement index and SARFII obtained to be 0.71875 and 0.9160045.

The system losses reduced to 1250.21 kW from 1473.1 kW with a net loss reduction of 222.88 kW and system losses got reduced by 15.13%.

Table 3.25: Optimal Allocation of solar DG for scenario1 under heavy load

Solar (Single DG Penetration)	Heavy load
Optimal Node (location)	Bus 18
Optimal size in kW	330
APL w/o DG (kW)	1473.1
APL W DG (kW)	1250.21
APLR (kW)	222.88
APLR (%)	15.13
Min V w/o DG (p.u.)	Bus 32
	0.75054

	Bus 9
Min V W DG (p.u)	0.84356
Voltage sag index before DG placement	0.9375
Number of sag buses (Nsag), Before DG placement	30
Number of sag buses (Nsag), After DG placement	23
Voltage sag index after DG placement	0.71875
SARFI,Before DG placement	0.0086207
SARFI,after DG placement	0.0078966
SARFII	0.9160045

Case-4, Scenario-2: optimal placement of Two solar DG units

Table 3.26 shows the results obtained from developed MSFLA algorithm for allocation of two Solar DG units for Heavy load condition. SARFI before installation of DG is obtained to be 0.0086207 and after DG integration it got reduced to 0.0077967, also SARFII obtained to be 0.9044161. The system losses reduced to 933.07 KW from 1473.1KW with a net loss reduction of 540.03 KW and system losses got reduced by 36.65 %. In the base case with heavy load condition the value of voltage sag index obtained to be 0.9375 and after integration of DG at the optimal location it reduced to 0.71875 and Voltage sag improvement index obtained to be 0.71875

Table 3.26: Optimal allocation of solar DG for scenario 2 under heavy load

Solar (Two DGs Penetration)	Heavy load
Optimal Node (location)	Bus 32 Bus 12
Optimal size in kW	993.9 1032
APL w/o DG (kW)	1473.1
APL W DG (kW)	933.07
APLR (kW)	540.03
APLR (%)	36.65
Min V w/o DG (p.u.)	Bus 32

	0.75054
Min V W DG (p.u)	Bus 9
	0.85702
Voltage sag index before DG placement	0.9375
Number of sag buses (Nsag), Before DG placement	30
Number of sag buses (Nsag), After DG placement	23
Voltage sag index after DG placement	0.71875
SARFI,Before DG placement	0.0086207
SARFI,after DG placement	0.0077967
SARFII	0.9044161

Case-5, Scenario-1: optimal placement of a wind Powered DG unit

Table 3.27 shows the results obtained from developed MSFLA algorithm for allocation of a Solar DG unit for Heavy load condition. Voltage sag index before (VSI) and SARFI before installation of DG is obtained to be 0.9375 and 0.0086207 and after DG integration it got reduced to 0.625 and 0.0068966, also Voltage sag improvement index and SARFII obtained to be 0.625 and 0.8000046.

The system losses reduced to 1101.28 KW from 1473.1kW with a net loss reduction of 371.81 kW and system losses got reduced by 25.24%.

Table 3.27: Optimal allocation of Wind powered DG for scenario1 under heavy load

Wind Power (Single DG Penetration)	**Heavy load**
Optimal Node (location)	Bus 31
Optimal size in	536 kW
	618.66 kVAR
APL w/o DG (kW)	1473.1
APL W DG (kW)	1101.28
APLR (kW)	371.81
APLR (%)	25.24
Min V w/o DG (p.u.)	Bus 32
	0.75054

Min V W DG (p.u)	Bus 9
	0.86213
Voltage sag index before DG placement	0.9375
Number of sag buses (Nsag), Before DG placement	30
Number of sag buses (Nsag), After DG placement	20
Voltage sag index after DG placement	0.625
SARFI,Before DG placement	0.0086207
SARFI,after DG placement	0.0068966
SARFII	0.8000046

Case-2, Scenario-2: optimal placement of two wind powered DG units

Table 3.28 shows the results obtained from developed MSFLA algorithm for allocation of two wind powered DG units for Heavy load condition. SARFI before installation of DG is obtained to be 0.0086207 and after DG integration it got reduced to 0.0054666, also SARFII obtained to be 0.6341242.

The system losses reduced to 878.07 KW from 1473.1kW with a net loss reduction of 595.03 kW and system losses got reduced by 40.39 %. In the base case with heavy load condition the value of voltage sag index obtained to be 0.9375 and after integration of DG at the optimal location it reduced to 0.50 and Voltage sag improvement index obtained to be 0.50.

Table 3.28: optimal allocation of wind power DG for scenario 2 under heavy load

Wind (Two DGs Penetration)	Heavy load
Optimal Node (location)	Bus 32
	Bus 12
Optimal size	506 kW, 608.66 kVAR
	511.53 kW, 597.85 kVAR
APL w/o DG (kW)	1473.1
APL W DG (kW)	878.07
APLR (kW)	595.03
APLR (%)	40.39
Min V w/o DG (p.u.)	Bus 32

	0.75054
Min V W DG (p.u)	Bus 9
	0.85229
Voltage sag index before DG placement	0.9375
Number of sag buses (Nsag), Before DG placement	30
Number of sag buses (Nsag), After DG placement	16
Voltage sag index after DG placement	0.50
SARFI,Before DG placement	0.0086207
SARFI,after DG placement	0.0054666
SARFII	0.6341242

Case-6, Scenario-1: optimal placement of solar and wind Powered DG unit

Table 3.29 shows the results obtained from developed MSFLA algorithm for allocation of a Solar DG unit for Heavy load condition. Voltage sag index before (VSI) and SARFI before installation of DG is obtained to be 0.9375 and 0.0086207 and after DG integration it got reduced to 0.375 and 0.0048789, also Voltage sag improvement index and SARFII obtained to be 0.375 and 0.5659517. The system losses reduced to 787.51 kW from 1473.1kW with a net loss reduction of 685.58 KW and system losses got reduced by 46.54%.

Table 3.29: optimal allocation of Solar and Wind powered DG under heavy load

Solar and Wind Power (DG Penetration)	Heavy load
Optimal Node (location)	Bus 06
	Bus 31
Optimal size in	309.21 kW
	512.21 kW ,602.12 Kvar
APL w/o DG (kW)	1473.1
APL W DG (kW)	787.51
APLR (kW)	685.58
APLR (%)	46.54
Min V w/o DG (p.u.)	Bus 32
	0.75054

Min V W DG (p.u)	Bus 9
	0.86321
Voltage sag index before DG placement	0.9375
Number of sag buses (Nsag), Before DG placement	30
Number of sag buses (Nsag), After DG placement	12
Voltage sag index after DG placement	0.375
SARFI,Before DG placement	0.0086207
SARFI,after DG placement	0.0048789
SARFII	0.5659517

3.5 SUMMARY:

In this section the optimal DG location and the penetration of renewable DGs for loss reduction and to improve voltage profile is determined through the logarithmic voltage deviation index method and modified shuffled frog leap algorithm.

 Then the system is subjected to heavy load condition and optimal allocation of DGs are done to prove the effectiveness of the method. Different indices are calculated to show the power quality improvement. In first method WSVPI is used to find the DG locations. In order to minimize the search space modified shuffled frog leap algorithm is new modified optimization technique which is used to find the optimum DG location. It has better convergence characteristics when compared to other algorithms.

The simulation results have indicated that the overall impact of the DG units on voltage profile is positive and proportionate reduction in power losses is achieved. It can be interfered that best results can be achieved with different renewable DGs.

4 Development of hybrid approach-based allocation of distributed generation to enhance system performance

4.1 INTRODUCTION

The intervention of the distributed generation has increased to a great extent in the power system. The optimal placement and optimal location for available penetration are the main concern to provide improved power quality in the network. Literature has indicated that any error in the selection of the location or size will lead to greater amount of losses as compared to that of the base case. In developing countries such as India, the quality of the power delivered to the customer through the utilities consists of many irregularities with higher losses and poor voltage profile. By the optimal allocation and optimal location for available penetration the utilities get an advantage of improved voltage regulation and also the system reliability is enhanced which will burden from the transmission and distribution system leading to a reduction in new investments.

Chapters 2 and 3 explained the analytical and the recent nature inspired modified shuffled frog leap algorithm which was used to allocate the DG with optimal location for fixed penetration level by using weighted sum of voltage index method, voltage profile index method and voltage sag, voltage sag reduction method. In this chapter where this problem is solved by formulating a multi-objective function based on reducing the logarithmic voltage deviation index (LVDI) and the real power losses of the system, the solution for which is obtained using the hybrid optimization techniques.

The problem is formulated into two objectives namely power losses and voltage profile improvement which is achieved by reducing the logarithmic voltage deviation index (LVDI). These two objectives are combined together to form a multi-objective function proposed in the research work.

The section 4.4 presents the comparison of the existing techniques and the proposed methodology to show the effective operation of the algorithm to solve the DG allocation and sizing problem.

The developed algorithm is developed in the MATLAB environment and is tested on the standard IEEE 33 Bus test system, standard IEEE 69 bus test system and KPTCL 32 bus feeder of Mysore, India.

The Particle swam optimization (PSO is an evolutionary calculation method that was suggested in 1995 both by Kennedy and Eberhart [13]. The originals PSO algorithm was inspired by bird blocks and schools of fish behavior. Assume that the research space is d-dimensional in the basic PSO technology. PSO is randomly initialized with a set of random molecules (solutions), then by generational update by searching for optima. For each iteration, the update of each particle is done by following two of the best values. The initial is the best solution (fitness) she has achieved yet. This is called a p_{best}.

The other best value followed by the particle spray optimizer is the better value, acquired by any particle of the population so far. This is the best global value and is called g_{best}..

The PSO is expressed mathematically as follows:

$$v_i^{k+1} = w * v_i^k + c_1 * r_1 * \left(Pbest_i - x_i^k\right) + c_2 * r_2 \times$$
$$(gbest - x_i^k) \tag{6}$$
$$x_i^{k+1} = x_i^k + v_i^{k+1} \tag{7}$$

Where i = 1,2, …, n and n is the population size, w is the weight of inertia, c_1 and c_2 are the acceleration constants in the interval [0, 1]. A particle's velocity and its position are updated and based on the new position; a recalculation of the objective function of the optimization problem is performed(x_i^{k+1}).

Part 1 of (6) provides PSO exploration capability. The second and third sections present respectively special thinking and cooperation particles. The PSO begins to hazardally place the particles in a problematic space. At for each iteration, the velocities of the particles are computed using (6). Once the velocities have been determined, the positions of the particles can be calculated using (7). The changing process of particle position continues until the final criterion is met.

4.2 Literature survey

The advancement of effectiveness of individual basic algorithms is the main goal of process of hybridization and in this way search space is expanded and the convergence enhances as well. this improves the local explorations also it becomes easier to design the effective algorithms which are flexible and coherent for the handling of problems regarding multi objective optimization. In the past few years, for HSFLA the various structures have been generated by the past research workers. In this section the suggested hybrid algorithms are highlighted and this is done by focusing on the specific benefits.

[126] suggested a new HSFLA for the solution of global optimization by the application of a strategy of adaptive learning in the Particle Swarm optimization.

For the basic SFLA the biggest disadvantage is the higher cost of computation, in memeplexes the self-learning potential is therefore likely to find best and faster local solutions. Hence, the speed of convergence increases by the algorithm and also it escapes the local optima. As per the outcomes of comparison, the suggested algorithm is effective to outperform other algorithms such as SFLA and PSO especially in terms of speed of convergence, local minima distance and for searching solutions the time needed for computation.

The convergence can be enhanced and the shuffled frog quality can be increased as well in the basic SFLA of the frogs shuffled. [127] introduced HSFLA modern method for the solution of various problems such as bi criteria permutation and scheduling the flow shop. A modern elite tabu search or ETS algorithm is also applied for the solutions of high quality generation. They also utilized the new approach for the revision of DIP which stands for the dynamic ideal point and employed pareto adaptive archive set to make use of solutions of non-dominated kinds when upper limit is obtained for the size of pareto archive.

To being trapped into neighborhood optima and increment decent variety of arrangement [123] proposed a HSFLA dependent on Chaos search. Turmoil calculation could improve xg which discontinues development by creating new ideal arrangements. To upgrade the combination rate, exactness, and arrangement strength in fundamental SFLA in tackling picture division issues, [122, 128] applied clonal choice Algorithm (CSA) to SFLA. CSA which is roused from insusceptible component of human body can supplant most horrendous antibodies in the populace with new irregular ones. In this manner, SFLA forms the most noticeably terrible subset of populace and CSA forms the best subset of populace.

The computational outcomes indicated that the proposed HSFLA outflanks K-implies calculation, CLONALG and basic SFLA.[129] applied another nearby inquiry dependent on Nelder–Mead (NM) calculation to essential SFLA in unraveling a nonlinear blended variable advancement issue. In the proposed HSFLA, the analysts endeavored to improve the nearby hunt capacity by supplanting the most noticeably terrible arrangement in the populace with another one which is superior to the present arrangement. Too, a meta-Lamarckian neighborhood search which is applied by [130] is a decent method to improve the accessibility in essential SFLA. [131] used a HSFLA by utilizing the essential thoughts of fake fish to improve the exactness of the arrangement. The trial results showed that the new HSFLA improves worldwide assembly and breaks nearby optima.

Additionally, as indicated by the assessment results, the proposed HSFLA could improve framework execution and controllability on the objective and furthermore decrease the client transmission power.

The DG's are small capacity generators that are introduced in the local distribution network [11]. It is required to understand the impacts of DG in order to avoid unforeseen circumstances [12]. It is observed that there is a loss of 14% occurring only in the distribution sector which can be brought down by the installation of DG's which can also control the power losses of the system [13]. The voltage profile improvement requires to be focused by optimally allocating the distributed generators in the system [14].

There are different types of DG's namely PV, wind, small hydro, fuel cell, biomass etc.,[11-16]. In the distribution sector, DG's are a boon for the utilities as well as for the customers especially where there is a deficiency in the transmission and distribution network [17]. There are many factors which are considered in the allocation of DG's such as technology used, number of units and their capacity, placement etc.,

Optimizations techniques are utilized to maximize the advantages posed by DG's as they give appropriate voltage profiles and lower the losses in the network [18-20]. The study has shown many heuristic techniques proposed to solve the optimization problems. The paper [28] shows that the constant load model determines the effect of different loads and sensitivity to frequency and voltage. The authors in the paper [22] present the application of dynamic programming for placement of DG taking into account the voltage improvement and loss reduction.

The study of the heuristic method was followed by the evolutionary techniques [13-15] to find the optimal location of the DG. The authors in [15] proposed an adaptive PSO (APSO) to place multiple DG's where the objective was to minimize the real power losses only. The most sensitive busses are identified for the placement of the DG in [16], same capacity DG is installed at all the buses and analytically calculates the lowest loss combination as the best size-site pair. A GA based algorithm is proposed in the paper [13] to install multiple DG's for cost minimization.

The PSO was developed through the modeling and execution of the simple social models inspired by a flock of birds and school of fishes [18]. It is considered to be recent findings of the combinational metaheuristic technique based on the population-based stochastic search algorithm [19, 20]. It has a two-step procedure i.e., calculating the velocity of the particle and updating its position. Hence, it requires less time for computation and less memory. The role of this algorithm is to find the optimal value to the problem depending on the nature of the animal society.

Until there is a stopping criteria satisfaction the process of shuffling and local search continues [104]. The individual memetics quality is enhanced by the memetic evolution. In SFLA the local quantity passes on among the individual various memetics and the memetics is transferred by the process of shuffling among the worldwide solutions [105]. For the worldwide research in large generation with larger memeplex size SFLA has shown to have a better performance and also higher time is needed for the computation. Along with this, the process of shuffling also increases the memetics quality on the basis of various cultures of memeplexes.

Between local and global search outcomes the information exchange ensures the robustness and flexibility of an algorithm [102]. In the alternate way, random and determinacy approaches are integrated in case of SFLA [105, 106]. The messages are exchanged successfully in the determinacy approach and the other form which is known as random approach ensures that the algorithm has robustness and flexibility. [105].

There are some other investigations regarding the SFLA alternatively to solve various multi objective problems regarding continuous, combinatorial optimization concerns with objective based functions. Some limitations are there in SFLA for the determination of the optimal solutions as well. In accordance to it, the research workers have been working for the basic SFLA enhancement. On the algorithm the enhancement is done by the SFLA (MSFLA) modification or by the hybridization process of both. Algorithmic improvement is made in case of MSFLA and this is done by the application of function replacement, heuristic and Meta heuristic functions. With other evolutionary algorithms the SFLA is integrated in case of HFSLA.

On the basis of outcomes from the researches in the past, in basic SFLA the evaluation phase requires a lot of modifications. The frog at the worst position will never jump for example on the best frog's position. The speed of convergence increases by this limitation and hence ends up in the premature convergence, in each memeplex the local search space is reduced and also there is a reduction in the probability of convergence.

A balance is tried to be obtained between a deep and wide exploration of the research space in basic SFLA which are considered to be closer to the optima. [107]. The change in leaping is so small that if worst and best frog's positions are close. Therefore, in the local optima it causes a trap and by several investigations premature convergence is reported and in basic SFLA it is regarded as a biggest concern. [108].

SFLA limitations are there and in order to overcome these limitations especially in the evolution phase the enhancement is needed in the process of evaluation as recommended by [102]. This is to make use of best information either local or global to get frog leaping and to solve the problems of multi objective optimizations to handle the non-dominated solutions they offered the algorithmic technique. [113] applied the sort based approach which was suggested by [114, 115] for the handling of problems regarding multi objectives. As per the outcomes of computation, the MSFLA presented is highly competitive and more beneficial to solve the complicated concerns especially when it is contrasted with the dynamic programming. it was also extended [109] that by employing an encoding based and urban based sequence on SFLA the new policy is employed as well to cope with the concerns regarding combinatorial optimization like salesman travelling. Effectiveness, diversity maintenance, convergence and time, cost of suggested MSFLA is indicated in the outcomes of the experiments.

In contrast to the problems of multi objective optimization, to handle such problems another algorithm is developed known as MSFLA [116, 117]. As fuzzy sets therefore the objective functions were formulated. For the improvement of SFLA and its stability and to find the global optima this method is employed and also it was suggested by [108] to determine for the selected frog the new position. CLS which stands for the chaotic local research is developed for the determination of newer solutions especially if the new frog quality was not improved in contrast with the worst frogs. On the basis of estimation outcomes, the algorithm suggested is accurate and in various applications it can be used as well. By the suggested MSFLA the economic load dispatch problem can be solved.

The behavior of local frogs in the swamps is simulated by the population-based algorithm known as SFLA since they are looking for the location with plenty of food. the local as well as global search potential is utilized by the algorithm [119] and also the MA and PSO algorithmic benefits are integrated with it. The information is passed among the population and it is the biggest benefit of MA, in case of GA only parent sibling interaction is seen. PSO algorithm on the other hand is more effective since it has faster rate of convergence.

To best global solutions its particles tend to converge and offer an optimal solution with higher convergence speed [105]. Along with this PSO algorithm unlike MS required a process of a simple coding and hence there are better benefits of MA and PSO due to better execution and it can determine the optimal solutions as well.

The SFLA population is meta heuristic in nature and solutions are contained in the form of frogs which on the basis of various frog cultures are classified into different memeplexes. Local exploration is done by SFLA making use of each memeplexes and algorithmic PSO at the same time. there is a shuffling and reorganization of frogs into the modern memeplexes and this is done to ensure the worldwide exploration [120].

In the MSFLA which is presented here, to hold the local non dominated solutions an external repository is defined. To manage the repository size another fuzzy clustering method is applied. It has been shown in the outcomes of MSFLA that it is more effective as compared to other used algorithms such as GO and PSO.

 Along with this the new mutation method is applied to reduce the time of computation and to increase the solution quality along with local optima trapping which normally takes place. Furthermore, [118] utilized a new method of MSFLA for the frog leaping modification and in this case the MSFLA tribe was used to prevent prematurity issues. As a tribe in MSFLA each memeplex is also considered as a tribe.

For the improvement of SFLA and its global searching potential, [121] employed a modern MSFLA and for the modification of SFLA division method the algorithm was suggested to balance the memeplexes performance. The new frog leaping rule is also used so that frogs can evolve more. It has been shown in the outcomes that as compared to SFLA and PSO algorithm the MSFLA are better. In this way for the basic SFLA enhancement and to make it more effective [122 and 124] applied MSFLA new by adding for each memeplex the local best value despite of developing a new frog.

With high quality of the population the initialization and search exploitation potential MSFLa was developed by [123].

In the MSFLA the following modifications were done at first, by making use of 3 initial rules and for the initialization of population, the second process in each memeplex is the evolution with the help of technique of crossing over for the formation of new solution. The third as well as local search is made with the help of neighboring infrastructures by the problem constraints consideration so that the problems can be solved. With the suggested MSFLA there is a large published algorithmic variety which can be used and these include PSO and SA hybrid algorithms, hybrid search PSO-tabu, AIA which stands for the artificial immune algorithm, MOGA which is an abbreviation of multi objective genetic algorithm and ACO (KBACO) which is knowledge-based algorithm. The effectiveness of the suggested algorithm is indicated by the outcomes of the experiments.

To improve the potential of exploration an effective and modern MSFLA was suggested by [120] and to solve the problems regarding multimode resource-based constraint the scheduling is done. In the suggested algorithm the population is generated by the improvement in forward backward direction and two-point crossing over is done by the suggested method so that encoded GA is scheduled traditionally. It has been shown in the outcomes of simulation that in the improvement of performance as compared to PSO the MSFLA has better working.

The novel MSFLA presented also has a potential to deal with the problems like flow shop scheduling. Insert neighborhood based local search is applied by the algorithm developed so that the potential of search increases.

The roulette wheel selection is also adapted to determine best frogs globally in the evolution's earlier stages so that the research is enhanced. MSFLA produce the outcomes which are better as compared to other algorithms such as SFLA and PSO.

The observations concluded that the MSFLA-PSO technique outperformed the other techniques by improving the accuracy and speed of operation by guiding the search toward a feasible solution.

The steps of execution of the algorithm are given as follows

Step1: Select the standard test system with all base case parameters

Step 2: Run load flow on the selected test system

Step 3: Select DG Wind and integrate in terms of Percentage penetration (viz: 10, 20, 30, 40, 50) Chose the optimal location by the hybrid algorithm for the selected penetration level and calculate PLRI and LVDI.

Step 4: Calculate WSOF using equation

Step 5: The obtained values of WSOF will gives the optimal location for with and without DG penetration.

Step 6 Integrate the DG with optimal location and size run load flow

Step 7: Calculate the power losses and WSOF for the optimal case of integration to show the improvement in voltage profiles and reduction of power losses.

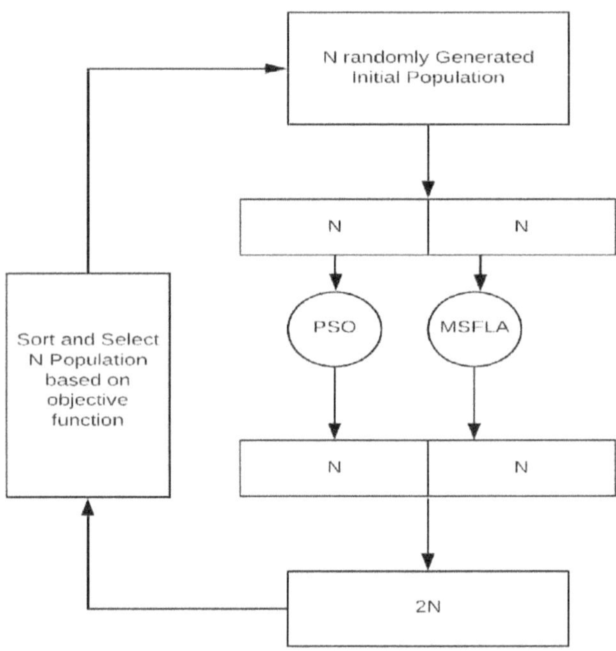

Fig 4.1: Schematic representation of proposed hybrid PSO and MSFLA.

4.3 DG Placement problem formulation

In this work, DG's are installed to enhance the system performance of the electrical network. Power quality can be improved by allocating DG at optimal location for particular penetration and further best results are obtained by optimal allocation of DG's using the hybrid MSFA-PSO. The inclusion of DG in an appropriate way will enhance the voltage profile of the network and also reduce the system losses, hence improving the overall performance of the system.

4.3.1 Power loss calculation

The equation of the real power losses in the system can be given in terms of exact loss formula as state below, the power loss is given by

$$P_L = \sum_{i=1, j=1}^{n} \left[\alpha_{ij}\left(P_i P_j + Q_i Q_j\right) + \beta_{ij}\left(Q_i P_j - P_i Q_j\right) \right] \dots\dots\dots\dots\dots\dots\dots\dots\dots\dots\dots\dots\,1$$

Where $\alpha_{ij} = \dfrac{r_{ij}}{V_i V_j} \mathbf{Cos}(\partial_i - \partial_j)$ and $\beta_{ij} = \dfrac{r_{ij}}{V_i V_j} \mathbf{Sin}(\partial_i - \partial_j)$

$$Q_L = \sum_{i=1, j=1}^{n} \left[\yen_{ij}\left(P_i P_j + Q_i Q_j\right) + \lambda_{ij}\left(Q_i P_j - P_i Q_j\right) \right] \dots\dots\dots\dots\dots\dots\dots\dots\dots\dots\dots\dots\,2$$

Where $\yen_{ij} = \dfrac{x_{ij}}{V_i V_j} \mathbf{Cos}(\partial_i - \partial_j)$ and $\lambda_{ij} = \dfrac{x_{ij}}{V_i V_j} \mathbf{Sin}(\partial_i - \partial_j)$

Penetration level: The penetration level of DG is chosen based on the following equation. .

$$PLDG\% = \frac{Size\,of\,DG}{Total\,demand} * 100 \quad \dots\dots\dots\dots\dots\dots\dots\dots\dots\dots\dots\dots\dots\dots\dots\dots\dots\dots\dots\,3$$

Where,

PLDG% is penetration Level of DG

The objective function for the minimization of the "real power loss and logarithmic voltage deviation index (LVDI)" can be expressed by the relation given by :

$$F_1 = PLRI = \frac{(P_{before} - P_{after})}{P_{before}} \dots\dots\dots\dots\dots\dots\dots\dots\dots\dots\dots\,(4.12)$$

Where, P_{loss1} and P_{loss2} are the power loss without and with DG.

$$F_2 = LVDI = \sum_{i=1}^{n} log\left(\frac{V_{i(new)}}{V_{i(old)}}\right) \quad where\,i = 1,2,3 \dots\dots n \dots\dots\dots\dots\dots\,(4.13)$$

Where, V_{old} is the voltage obtained after performing the load flow and V_{new} is the value of the voltage after placement of DG.

The main objective function is to minimize F in such a way that;

$$\textit{Minimize } \mathbf{F = [(w1*F_1) + (w2*F_2)]} \,\dots\dots\dots\dots\dots\,\mathbf{(4.14)}$$

For optimal allocation the size and the Location of DG units will be decided by the algorithm based on the multi-objective function considering power loss reduction and logarithmic voltage deviation with respect to the weighted sum of w1 and w2 equal to unity. The weightage can be decided based on the analysis with the importance given to the parameters (voltage or losses).

For the selected penetration level optimal location will be chosen by the algorithm based on objective function.

Where, F_1 is the power loss minimization index and F_2 is the voltage profile improvement by reducing the deviation index. The set of equality and inequality constraints are,

$$P_{Gj}^{min} \leq P_{Gj} \leq P_{Gj}^{max} \dots \dots \dots \dots j = 1, 2, \dots . Ng$$

$$Q_{Gj}^{min} \leq Q_{Gj} \leq Q_{Gj}^{max} \dots \dots \dots \dots j = 1, 2, \dots . Ng$$

$$P_{DG}^{min} \leq P_{DG} \leq P_{DG}^{max}$$

$$V_{n}^{min} \leq V_{n} \leq V_{n}^{max} \dots \dots \dots \dots n = 1, 2, \dots . Nd$$

4.4 Results and discussions

In this part, the results obtained after the execution of a new shuffled frog leap algorithm and particle swarm optimization algorithm is done on the IEEE 33, IEEE69 and Practical 32 bus system of Mysuru, Karnataka, India in MATLAB.

For each test system, the following six cases are considered and for the first two cases, in each case two scenarios are considered.

Case -1 : Allocation of Solar DG unit

Case -2 : Allocation of Wind DG unit

Case -3 : Allocation of Hybrid DG unit

Scenario 1 : 10 % Penetration of DG

Scenario 2 : 20 % Penetration of DG

Scenario 3 : 30 % Penetration of DG

Scenario 4 : 40 % Penetration of DG

Scenario 5 : 50 % Penetration of DG

Case -4 : Optimal Allocation of Solar DG unit

Case -5 : Optimal Allocation of Wind DG unit

Scenario 1 : Placement of single DG

Scenario 2 : Placement of Two DGs

Case -6 : Optimal Allocation of Hybrid DG unit

4.4.1 CASE 1: IEEE 33 BUS TEST SYSTEM

The developed method is applied to the standard IEEE 33 bus system operated at 12.66kV level of voltage and consists of 33 busses 32 branches. The line and load data are available in the appendix A. 3715kW and 2300kVar is the total active and reactive power of the system respectively. Fig 4.2shows the network considered. The base case load flow is obtained where the power losses without the connection of the DG's are 202.68kW and 143.22kVAr respectively.

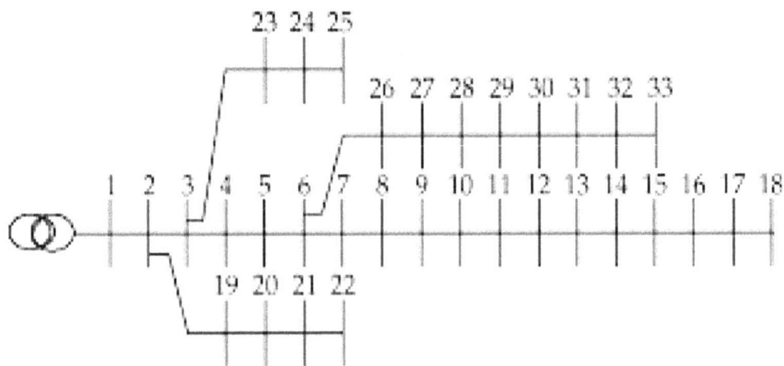

Fig 4.2: Single line diagram of IEEE 33 bus

Particulars	**Abbreviations**
1: Optimal Location of DG	OLDG
2: Optimal DG Size	ODGS
3: Base Case Power Losses (kW)	BCPL (kW)
4: Power Loss with DG (kW)	PL w DG (kW)
5: Power Loss Reduction in (%)	PLR (%)

Case-1, Scenario-1: 10 % integration of a SOLAR DG

The results from developed MSFA-PSO algorithm for allocation of a Solar DG unit for scenario-1 are tabulated in table 4.1. The system losses reduced to 161.01 kW from 202.68kW and percentage losses are reduced to 20.55. Minimum voltage before DG placement is 0.91309 p.u at bus 18 and after DG placement the minimum voltage obtained to be 0.92258 p.u.at bus 33.

Figure 4.3 shows the variation of voltage and power losses at different buses before and after integration of 10% of Solar DG.

Table 4.1: Allocation of Solar DG for scenario 1

Particulars	MSFA 10% SOLAR DG	MSFA-PSO 10% SOLAR DG
1: OLDG	Bus 06	Bus 06
2: ODGS	371.5 kW	371.5 kW
3: BCPL (kW)	202.68	202.68
4: PLDG (kW)	166.2	161.01
5: PLR (%)	17.99	20.55
6: Min V w/o DG (p.u.)	Bus 18 0.91309	Bus 18 0.91309
7: Min V W DG (p.u.)	Bus 18 0.92056	Bus 33 0.92258

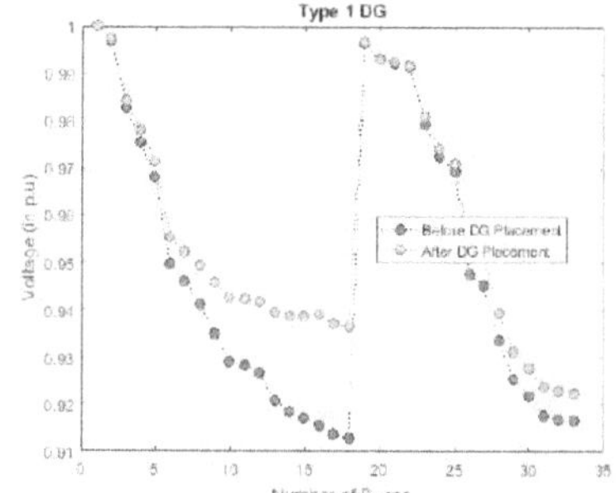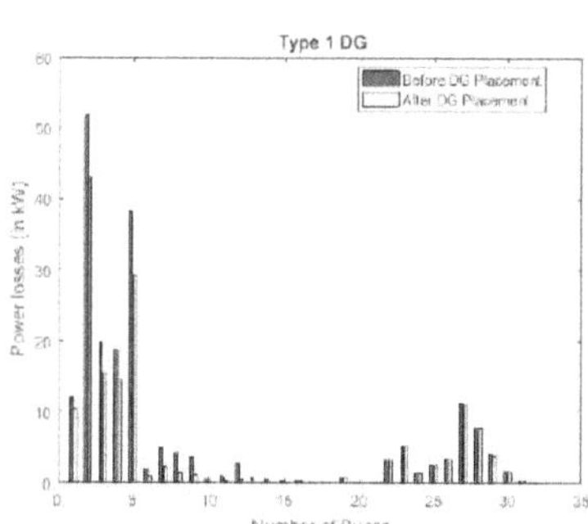

Fig. 4.3: voltage profile and power losses of the IEEE33 bus system for integration of Type1 DG at 10% Penetration.

Case-1, Scenario-2: 20 % integration of a SOLAR DG

The results from developed MMSFA-PSO   algorithm for allocation of a Solar DG unit for scenario-2 are tabulated in table 4.2. The system losses reduced to 137.65 kW from 202.68kW and percentage losses are reduced to 32.08. Minimum voltage before DG placement is 0.91309 p.u at bus 18 and after DG placement the minimum voltage obtained to be 0.9282 p.u.at bus 33.

Table 4.2: Allocation of Solar DG for scenario 2

Particulars	MSFA 20% SOLAR DG	MSFA-PSO 20% SOLAR DG
1: OLDG	Bus 16	Bus 14
2: ODGS	743kW	743 kW
3: BCPL (kW)	202.68	202.68
4: PLDG (kW)	145.01	137.65
5: PLR (%)	28.45	32.08
6: Min V w/o DG (p.u.)	Bus 18	Bus 18
	0.91309	0.91309
7: Min V W DG (p.u.)	Bus 33	Bus 33
	0.92091	0.9282

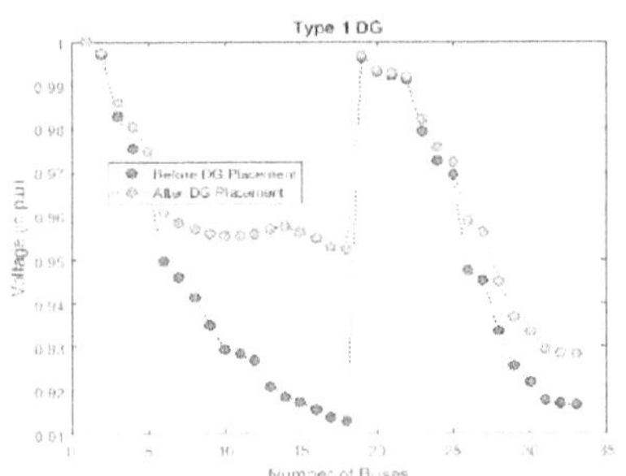
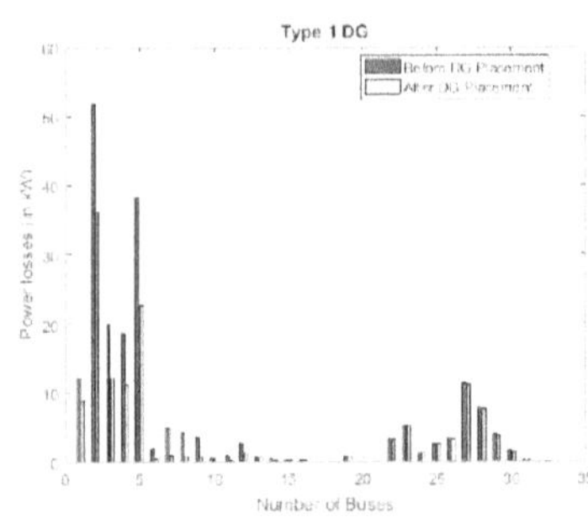

Fig. 4.3(a): voltage profile and power losses of the IEEE33 bus system for integration of
Type1 DG at 20% Penetration.

Figure 4.3(a) shows the variation of voltage and power losses at different buses before and after
integration of 20% of Solar DG.

Case-1, Scenario-3: 30 % integration of a SOLAR DG

The results from developed MMSFA-PSO   algorithm for allocation of a Solar DG unit for
scenario-3 are tabulated in table 4.3. The system losses reduced to 123.56 kW from 202.68kW
and percentage losses are reduced to 39.03. Minimum voltage before DG placement is 0.91309
p.u at bus 18 and after DG placement the minimum voltage obtained to be 0.9302 p.u.at bus

114

18. Figure 4.4 shows the variation of voltage and power losses at different buses before and after integration of 30% of Solar DG.

Table 4.3: Allocation of Solar DG for scenario 3

Particulars	MSFA 30% SOLAR DG	MSFA-PSO 30% SOLAR DG
1: OLDG	Bus 6	Bus 30
2: ODGS	1114.5 kW	1114.5 kW
3: BCPL (kW)	202.68	202.68
4: PLDG (kW)	123.71	120.56
5: PLR (%)	38.96	39.03
6: Min V w/o DG (p.u.)	Bus 18	Bus 18
	0.91309	0.91309
7: Min V W DG (p.u.)	Bus 18	Bus 33
	0.92921	0.9302

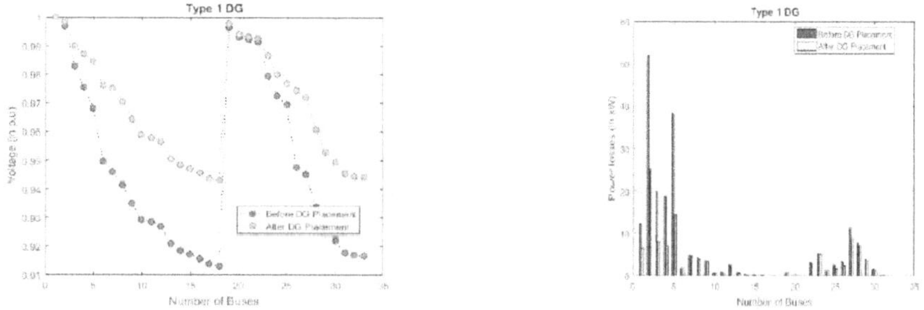

Fig. 4.4: voltage profile and power losses of the IEEE33 bus system for integration of Type1 DG at 30% Penetration.

Case-1, Scenario-4: 40 % integration of a SOLAR DG

The results from developed MMSFA-PSO   algorithm for allocation of a Solar DG unit for scenario-4 are tabulated in table 4.4. The system losses reduced to 116.51 KW from 202.68KW and percentage losses are reduced to 42.51%. Minimum voltage before DG placement is 0.91309 p.u at bus 18 and after DG placement the minimum voltage obtained to be 0.93552 p.u.at bus 18. Figure 4.5 shows the variation of voltage and power losses at different buses before and after integration of 40% of Solar DG.

Table 4.4: Allocation of Solar DG for scenario 4

Particulars	MSFA 40% SOLAR DG	MSFA-PSO 40% SOLAR DG
1: OLDG	Bus 30	Bus 14
2: ODGS	1486 kW	1486 kW
3: BCPL (kW)	202.68	202.68
4: PLDG (kW)	120.30	116.51
5: PLR (%)	40.64	42.51
6: Min V w/o DG (p.u.)	Bus 18	Bus 18
	0.91309	0.91309
7: Min V W DG (p.u.)	Bus	Bus 18
	0.93325	0.93552

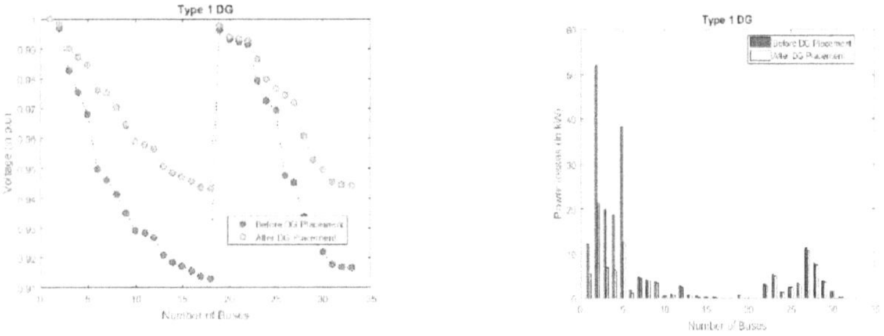

Fig. 4.5: voltage profile and power losses of the system for integration of Type1 DG at 40% Penetration.

Case-1, Scenario-5: 50 % integration of a SOLAR DG

The results from developed MMSFA-PSO   algorithm for allocation of a Solar DG unit for scenario-5 are tabulated in table 4.5. The system losses reduced to 110.24 kW from 202.68 kW and percentage losses are reduced to 45.6. Minimum voltage before DG placement is 0.91309 p.u at bus 18 and after DG placement the minimum voltage obtained to be 0.94318 p.u.at bus

18. Figure 4.6 shows the variation of voltage and power losses at different buses before and after integration of 50% of Solar DG.

Table 4.5: Allocation of Solar DG for scenario 5

Particulars	MSFA 50% SOLAR DG	MSFA-PSO 50% SOLAR DG
1: OLDG	Bus 27	Bus 14
2: ODGS	1857.5 kW	1857.5 kW
3: BCPL (kW)	202.68	202.68
4: PLDG (kW)	113.4	110.24
5: PLR (%)	44.05	45.6
6: Min V w/o DG (p.u.)	Bus 18 0.91309	Bus 18 0.91309
7: Min V W DG (p.u.)	Bus 18 (0.94083pu)	Bus 18 (0.94318pu)

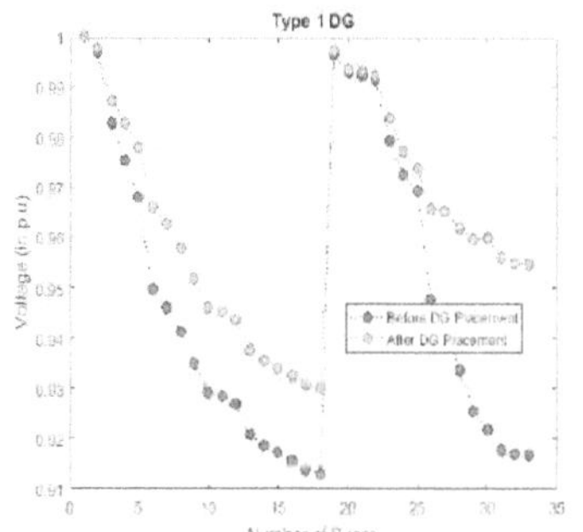
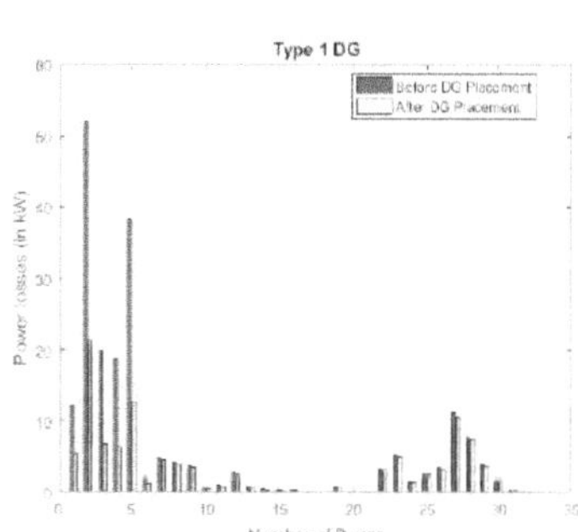

Fig. 4.6: voltage profile and power losses of the test system for integration of Type1 DG at 50% Penetration

Case-1, Scenario 1- : optimal placement of a SOLAR DG

The results obtained from the developed MMSFA-PSO   algorithm for case-1, scenario-1 and comparison with SFA is tabulated in table.4.6. Total losses of the system are reduced to 103.97 kW from 202.68 kW with percentage losses are reduced to 48.70%.

Table 4.6: Optimal Allocation of Solar DG by MMSFA-PSO and compared with MSFA

Particulars	MSFA SOLAR DG	MSFA-PSO SOLAR DG
1: OLDG	Bus 7	Bus 06
2: ODGS (kW)	2315.6	2593.3
3: BCPL (kW)	202.68	202.68
4: PLDG (kW)	102.8	103.97
5: PLR (%)	50.72	48.70
6: Min V w/o DG (p.u.)	Bus 18 0.91309	Bus 18 0.91309
7: Min V W DG (p.u.)	Bus 18 0.97526	Bus 18 0.98131

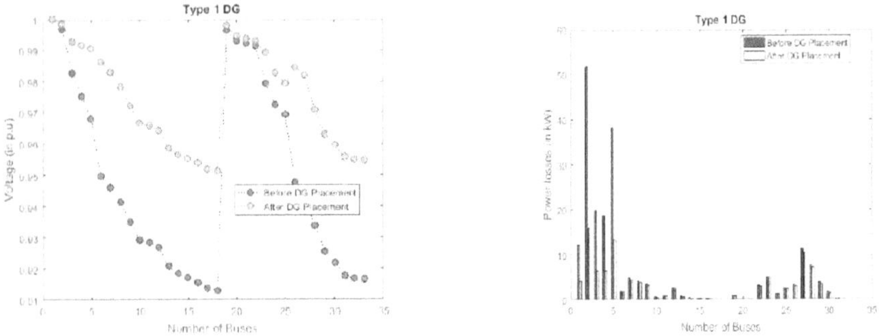

Fig. 4.7: voltage profile and power losses of the IEEE33 bus system for optimal integration of Type1 DG.

Case-1, Scenario-2: optimal placement of Two SOLAR DG units

The results obtained from the developed MMSFA-PSO   algorithm for case-1, scenario-2 is tabulated in table.4.7. Total losses of the system are reduced to 87.88KW from 202.68KW.

Table 4.7: Optimal allocation of two Solar DG by MSFA-PSO

Particulars	MSFA-PSO SOLAR DG
1: OLDG	Bus 30, 12
2: ODGS (kW)	1253.5 1109.2
3: BCPL (kW)	202.68
4: PLDG (kW)	87.88
5: PLR (%)	56.63
6: Min V w/o DG (p.u.)	Bus 18 0.91309
7: Min V W DG (p.u.)	Bus 18 0.97402

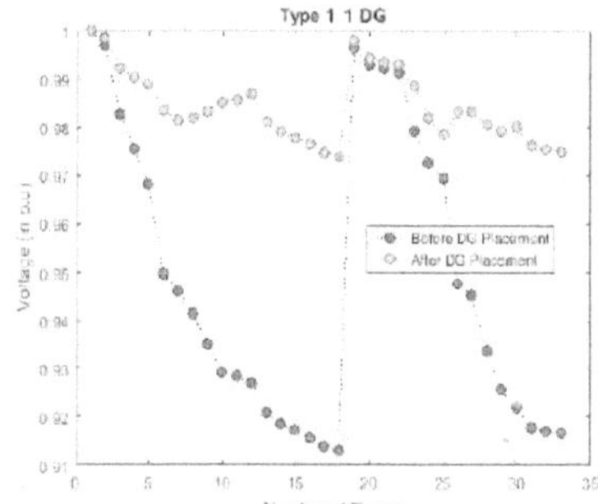

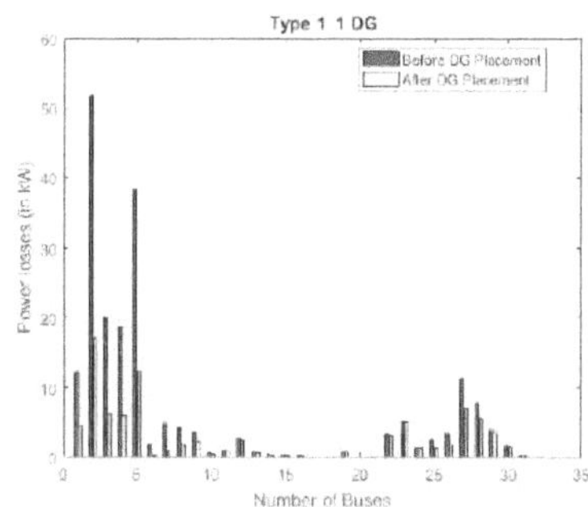

Fig. 4.8: voltage profile and power losses of the IEEE33 bus system for two optimal integration of Type1 DGs

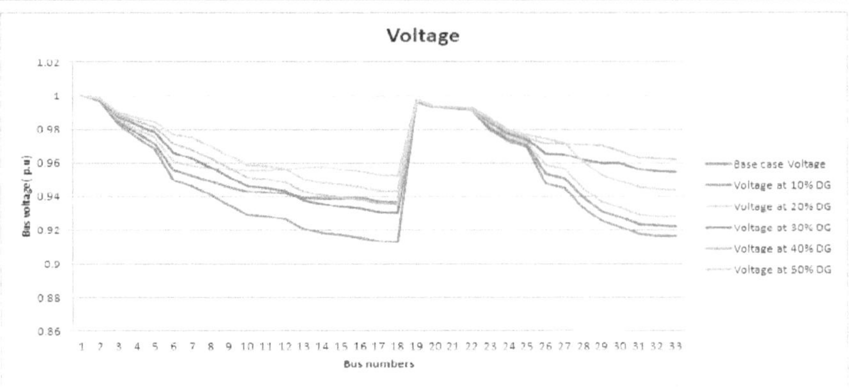

Figure 4.9: Voltage at different buses with and without Solar DG penetration

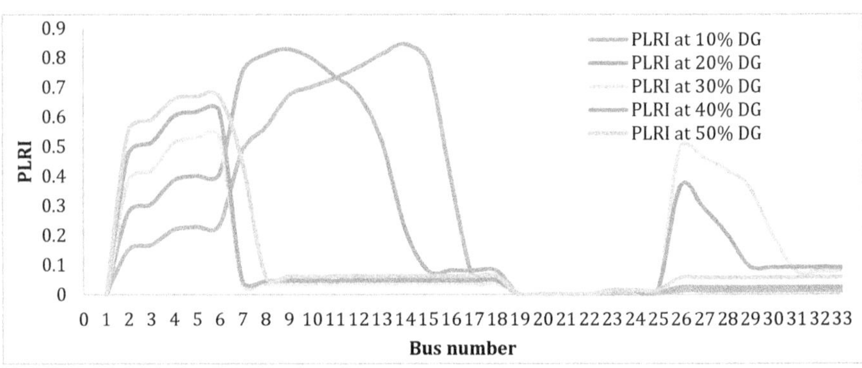

Figure 4.10: PLRI at different buses with Solar DG penetration at different levels

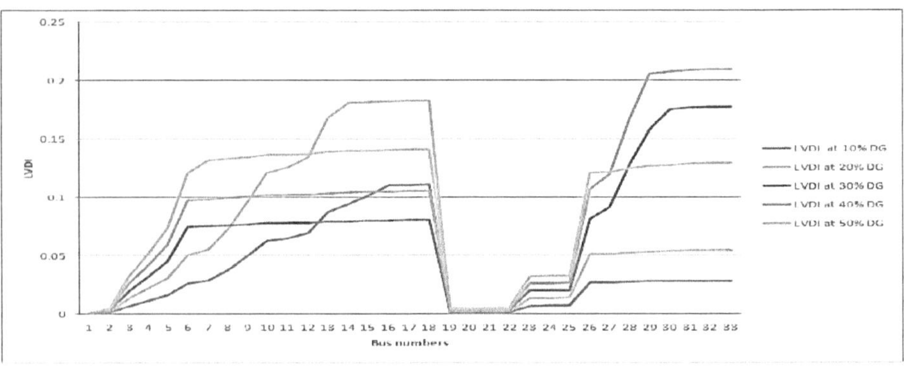

Figure 4.11: LVDI at different buses with Solar DG penetration at different levels

Table4.8 : WSOF values at different buses after integration Solar DG at different penetration levels

Bus number	WSOF at 10%	WSOF at 20%	WSOF at 30%	WSOF at 40%	WSOF at 50%
1	0	0	0	0	0
2	0.060915	0.111963	0.155443	0.192384	0.224063
3	0.070877	0.129678	0.179081	0.220363	0.254963
4	0.094147	0.167542	0.224219	0.266083	0.295152
5	0.100901	0.17905	0.23875	0.282151	0.311345
6	0.10889	0.193631	0.258868	0.306848	0.339987
7	0.210927	0.33294	0.059159	0.077276	0.258921
8	0.248199	0.367268	0.059728	0.078029	0.104065
9	0.298676	0.388989	0.060461	0.079002	0.105356
10	0.318372	0.391335	0.061096	0.079802	0.106435
11	0.331563	0.369884	0.06122	0.079953	0.106641
12	0.348918	0.350614	0.061405	0.080196	0.106923
13	0.376819	0.309519	0.06205	0.081032	0.108042
14	0.394497	0.199035	0.062304	0.08138	0.1085
15	0.374425	0.141233	0.062467	0.081596	0.108787
16	0.222606	0.141609	0.062651	0.081806	0.109095
17	0.086094	0.142096	0.06285	0.0821	0.109448
18	0.086213	0.142244	0.06295	0.082192	0.109599
19	0.000852	0.001628	0.002403	0.003125	0.003899
20	0.000832	0.001649	0.002404	0.003158	0.003913
21	0.000855	0.001657	0.002459	0.003195	0.003931
22	0.000849	0.001646	0.002434	0.003185	0.003937
23	0.005469	0.010617	0.015576	0.020387	0.025231
24	0.005528	0.010731	0.015748	0.020622	0.025509
25	0.005549	0.010787	0.01585	0.02077	0.025653
26	0.020977	0.04048	0.248188	0.209458	0.095364
27	0.02106	0.040673	0.240406	0.188512	0.095819
28	0.021462	0.041469	0.245469	0.182777	0.097642

29	0.021775	0.042044	0.239429	0.161726	0.09898
30	0.021907	0.042302	0.187102	0.160926	0.099602
31	0.022094	0.042643	0.137504	0.16218	0.100384
32	0.022147	0.042732	0.137762	0.162483	0.100571
33	0.022168	0.042761	0.1378	0.16257	0.100614

Case-2, Scenario-1: 10 % integration of a wind DG

The results from developed MSFA-PSO algorithm for allocation of a wind DG unit for scenario-1 are tabulated in table below. the analysis is performed for three different cases which is further compared to the previously calculated value of the modified shuffled frog leap algorithm. The first technique allocates the DG according to the fixed value of active and reactive power of the introduced wind generator in the system. Here, the system losses are reduced to 143.8 kW from 202.68 kW and percentage losses are reduced to 29.04. the next technique allocates the DG according to the fixed value of active power and optimal value of the reactive power which is calculated by the algorithm. here, the system losses reduced are to 143.8 kW from 202.68 kW and percentage losses are reduced to 29.04. the last approach allocates the DG according to the optimal value of the active and reactive power which is calculated by the algorithm. here, the system losses reduced are to 143.8 kW from 202.68 kW and percentage losses are reduced to 29.04. Minimum voltage before DG placement is 0.91309 p.u at bus 18 and after DG placement the minimum voltage obtained from the best of the three approaches is observed to be 0.92149 p.u.at bus 18. Figures shows the graphs of voltage and power losses at different buses before and after integration of 10% of wind DG.

Table 4.9: Allocation of wind DG for scenario 1

Particulars	MSFA 10% wind DG	MSFA-PSO 10% wind DG (fixed P, Q)	MSFA-PSO 10% wind DG (Fixed P, Optimal Q within 10%)	MSFA-PSO wind DG (Optimal P, Q within 10%)
1: OLDG	Bus 6	Bus 32	Bus 14	Bus 32
2: ODGS (kW)	371.5 kW 230 kVAR	371.5 kW 230 kVAR	371.5 kW -136.24 kVAR	304.76 kW -214.94 kVAR
3: BCPL (kW)	202.68	202.68	202.68	202.68

4: PLDG (kW)	145.12	143.8	146.23	150.15
5: PLR (%)	28.39	29.04	27.85	25.91
6: Min V w/o DG (p.u.)	Bus 18 0.91309	Bus 18 0.91213	Bus 18 0.91321	Bus 18 0.91293
7: Min V W DG (p.u.)	Bus 18 0.92011	Bus 18 0.92149	Bus 18 0.92483	Bus 18 0.92093

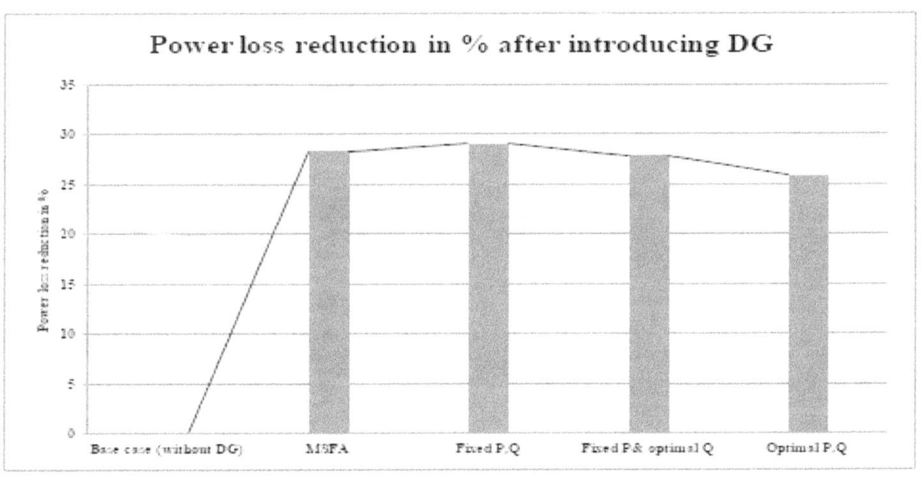

Figure 4.12: Voltage profile and power losses of the test system for wind integration for scenario 1

Figure 4.12: Power loss reduction in % after introducing DG for MSFA and scenario 1 of MSFA-PSO

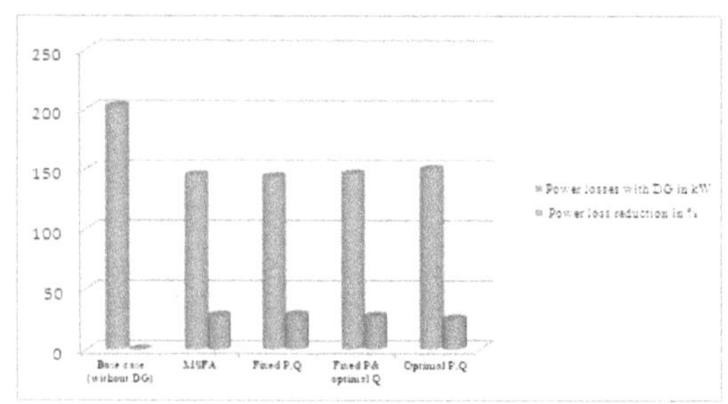
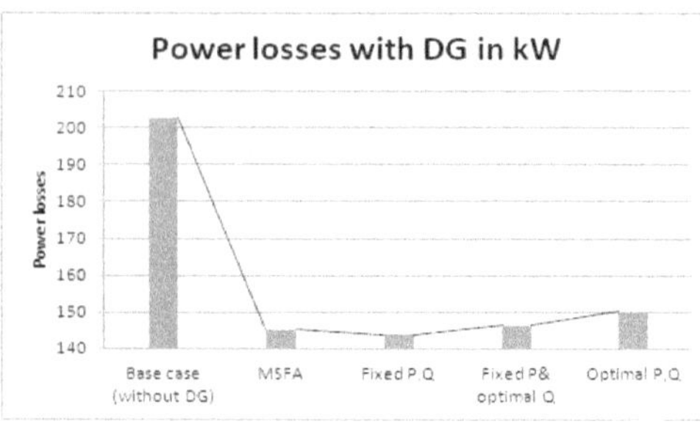

Figure 4.13: Power loss reduction in kW & % after introducing DG for MSFA and scenario 1

Case-2, Scenario-2: 20 % integration of a wind DG

The results from developed MSFA-PSO   algorithm for allocation of a wind DG unit for scenario-2 are tabulated in table below. the analysis is performed for three different cases which is further compared to the previously calculated value of the modified shuffled frog leap algorithm. The first technique allocates the DG according to the fixed value of active and reactive power of the introduced wind generator in the system. Here, the system losses are reduced to 105.08 kW from 202.68kW and percentage losses are reduced to 48.02. the next technique allocates the DG according to the fixed value of active power and optimal value of the reactive power which is calculated by the algorithm. here, the system losses reduced are to 128.07 kW from 202.68kW and percentage losses are reduced to 36.81. The last approach allocates the DG according to the optimal value of the active and reactive power which is calculated by the algorithm. here, the system losses reduced are to 125.55 kW from 202.68 kW and percentage losses are reduced to 38.05. Minimum voltage before DG placement is 0.91309 p.u at bus 18 and after DG placement the minimum voltage obtained from the best of the three approaches is observed to be 0.93031 p.u.at bus 33. Figures show the graphs of voltage and power losses at different buses before and after integration of 20% of wind DG.

Table 4.10: Allocation of wind DG for scenario 2

Particulars	MSFA 20% wind DG	MSFA-PSO 20% wind DG (fixed P, Q)	MMSFA-PSO 20% wind DG (Fixed P, Optimal Q within 20%)	MMSFA-PSO wind DG (Optimal P, Q within 20%)
1: OLDG	Bus 33	Bus 33	Bus 11	Bus 12
2: ODGS	743 kW 460 kVAR	743 kW -460 kVAR	743 kW -162.48 kVAR	715.15 kW -221.25 kVAR

3: BCPL (kW)	202.68	202.68	202.68	202.68
4: PLDG (kW)	139.01	105.08	128.07	125.55
5: PLR (%)	31.41	48.02	36.812	38.05
6: Min V w/o DG (p.u.)	Bus 18	Bus 18	Bus 18	Bus 18
	0.91309	0.91309	0.91309	0.91309
7: Min V W DG (p.u.)	Bus 18	Bus 18	Bus 18	Bus 33
	0.92531	0.9298	0.93009	0.93031

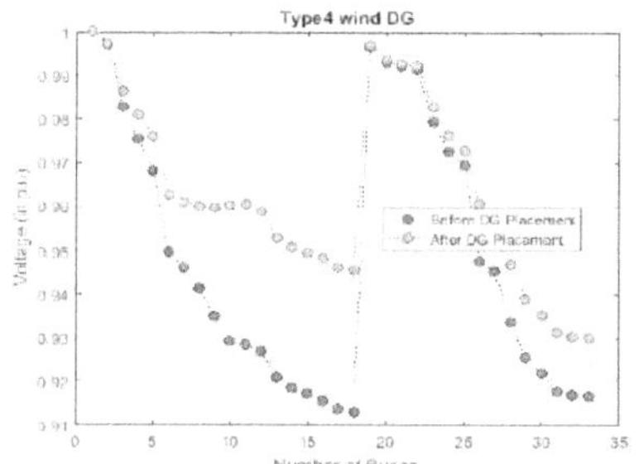
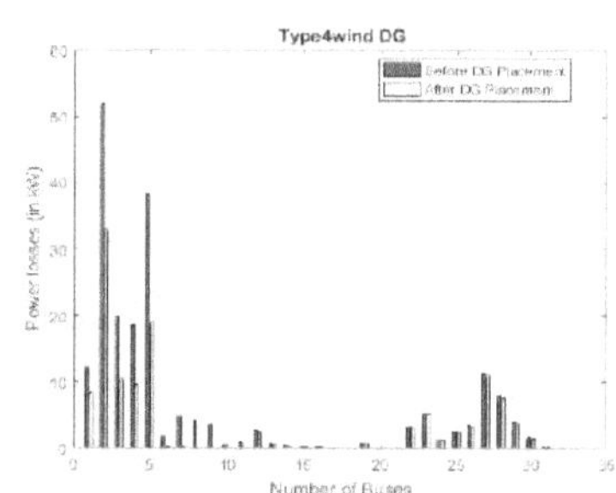

Figure 4.14: Voltage profile and power losses of the test system for wind integration for scenario 2

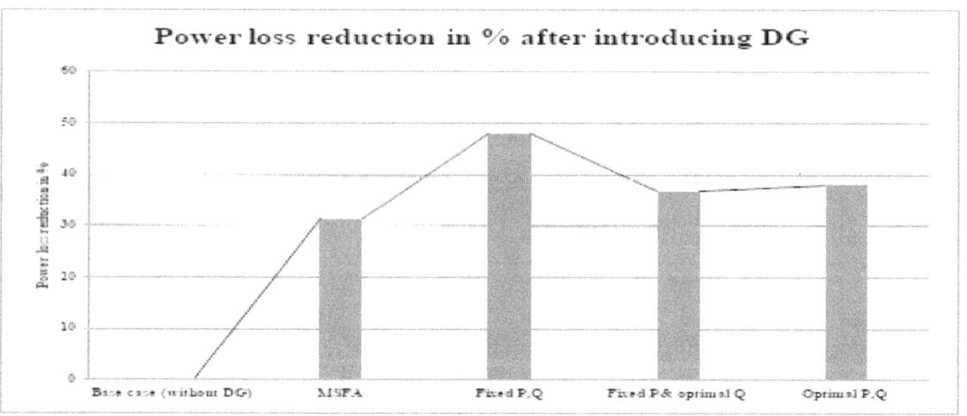

Figure 4.15: Power loss reduction in % after introducing DG for MSFA and scenario 2 of MSFA-PSO

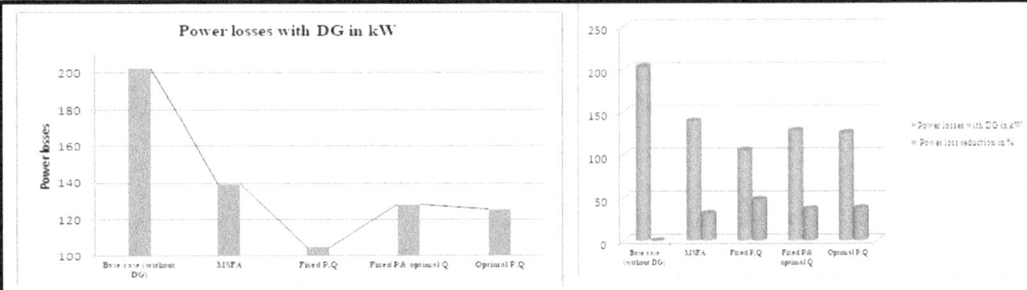

Figure 4.16: Power loss reduction in kW & % after introducing DG for MSFA and scenario 2 of MSFA-PSO

Case-2, Scenario-3: 30 % integration of a wind DG

The results from developed .MSFA-PSO algorithm for allocation of a wind DG unit for scenario-3 are tabulated in table below. the analysis is performed for three different cases which is further compared to the previously calculated value of the modified shuffled frog leap algorithm. The first technique allocates the DG according to the fixed value of active and reactive power of the introduced wind generator in the system. Here, the system losses are reduced to 95.84 kW from 202.68 kW and percentage losses are reduced to 52.27. The next technique allocates the DG according to the fixed value of active power and optimal value of the reactive power which is calculated by the algorithm. here, the system losses reduced are to 98.85 kW from 202.68 kW and percentage losses are reduced to 51.22. the last approach allocates the DG according to the optimal value of the active and reactive power which is calculated by the algorithm. here, the system losses reduced are to 96.12 kW from 202.68 kW and percentage losses are reduced to 52.57. Minimum voltage before DG placement is 0.91309 p.u at bus 18 and after DG placement the minimum voltage obtained from the best of the three approaches is observed to be 0.94954 p.u.at bus 33. Figures shows the graphs of voltage and power losses at different buses before and after integration of 30% of wind DG.

Table 4.11: Allocation of wind DG for scenario 3

Particulars	MSFA 30% wind DG	MSFA-PSO 30% wind DG (fixed P, Q)	MSFA-PSO 30% wind DG (Fixed P, Optimal Q within 30%)	MSFA-PSO wind DG (Optimal P , Q within 30%)
1: OLDG	Bus 33	Bus 11	Bus 11	Bus 08

2: ODGS	1145 kW 690 kVAR	1145 kW -690 kVAR	1145 kW -532.32 kVAR	1085.2 kW -565.42 kVAR
3: BCPL (kW)	202.68	202.68	202.68	202.68
4: PLDG (kW)	119.58	95.84	98.85	96.12
5: PLR (%)	41.39	52.71	51.226	52.57
6: Min V w/o DG (p.u.)	Bus 18 0.91309	Bus 18 0.91309	Bus 18 0.91309	Bus 18 0.91309
7: Min V W DG (p.u.)	Bus 18 0.92691	Bus 33 0.94097	Bus 33 0.93915	Bus 33 0.94954

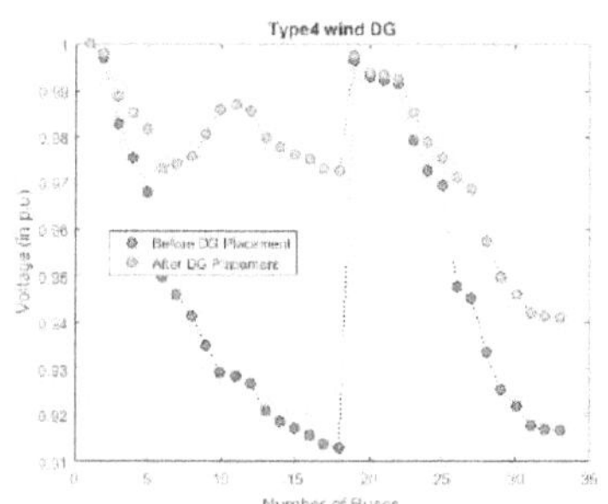
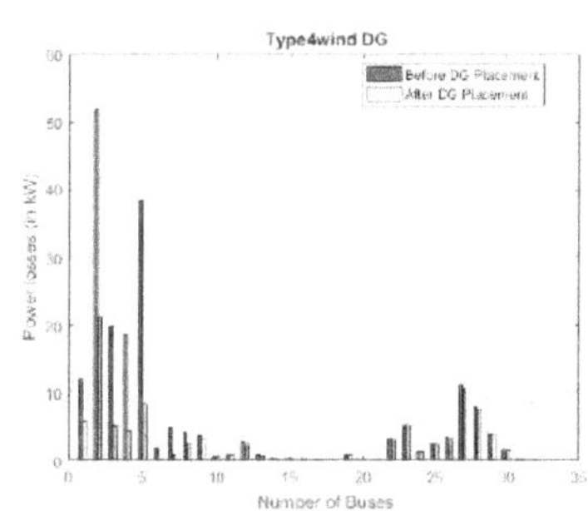

Figure 4.17: Voltage profile and power losses of the test system for wind integration for scenario 3

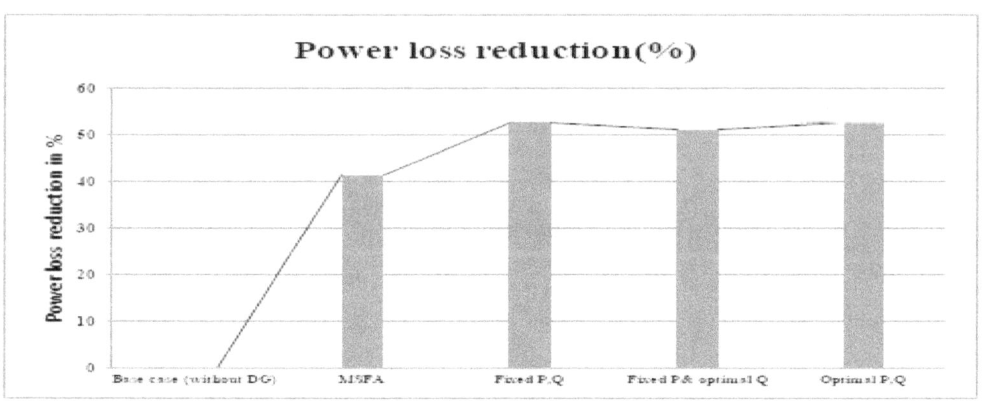

Figure 4.18: Power loss reduction in % after introducing DG for MSFA and scenario 3 of MSFA-PSO

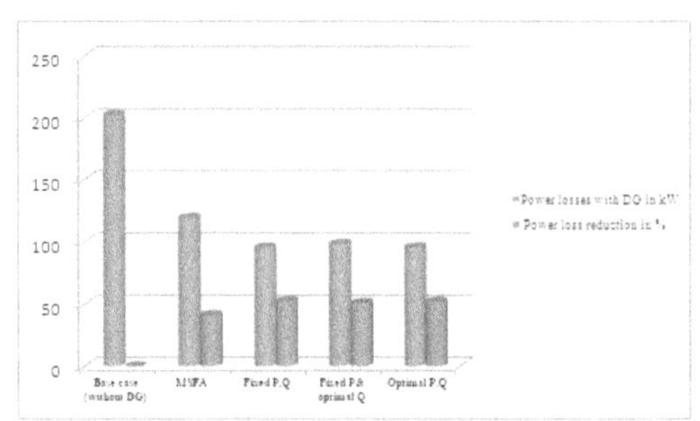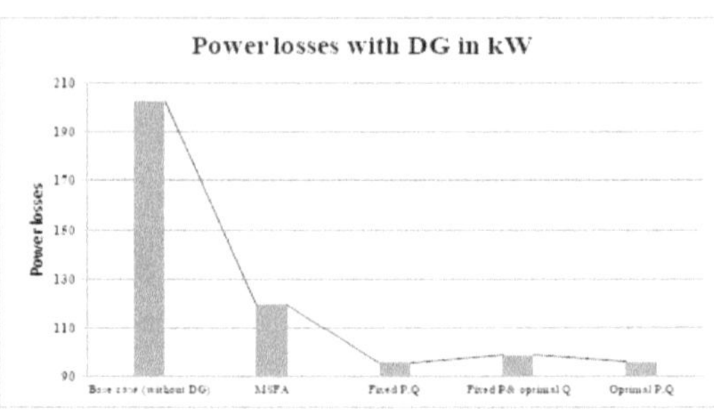

Figure 4.19 : Power loss reduction in kW & % after introducing DG for MSFA and scenario 3of MSFA-PSO

Case-2, Scenario-4: 40 % integration of a wind DG

The results from developed MSFA-PSO algorithm for allocation of a wind DG unit for scenario-1 are tabulated in table below. The analysis is performed for three different cases which is further compared to the previously calculated value of the modified shuffled frog leap algorithm. The first technique allocates the DG according to the fixed value of active and reactive power of the introduced wind generator in the system. Here, the system losses are reduced to 99.84 kW from 202.68 kW and percentage losses are reduced to 50.73. The next technique allocates the DG according to the fixed value of active power and optimal value of the reactive power which is calculated by the algorithm. Here, the system losses reduced are to 103.32 kW from 202.68 kW and percentage losses are reduced to 49.02. The last approach allocates the DG according to the optimal value of the active and reactive power which is calculated by the algorithm. Here, the system losses reduced are to 86.1 kW from 202.68kW and percentage losses are reduced to 57.51. Minimum voltage before DG placement is 0.91309 p.u at bus 18 and after DG placement the minimum voltage obtained from the best of the three approaches is observed to be 0.95023 p.u.at bus 18. Figures shows the graphs of voltage and power losses at different buses before and after integration of 40% of wind DG

Table 4.12: Allocation of wind DG for scenario 4

Particulars	MSFA 40% wind DG	MSFA-PSO 40% wind DG (fixed P, Q)	MSFA-PSO 40% wind DG (Fixed P, Optimal Q within 40%)	MSFA-PSO wind DG (Optimal P, Q within 40%)
1: OLDG	Bus 8	Bus 06	Bus 26	Bus 28
2: ODGS	1486 kW	1486 kW	1486 kW	1390 kW
	920 kVAR	-920 kVAR	-403.5 kVAR	-214.6 kVAR

3: BCPL (kW)	202.68	202.68	202.68	202.68
4: PLDG (kW)	121.31	99.84	103.32	86.1
5: PLR (%)	40.14	50.73	49.02	57.51
6: Min V w/o DG (p.u.)	Bus 18	Bus 18	Bus 18	Bus 18
	0.91309	0.91309	0.91309	0.91309
7: Min V W DG (p.u.)	Bus 18	Bus 18	Bus 18	Bus 18
	0.93953	0.94461	0.93961	0.95023

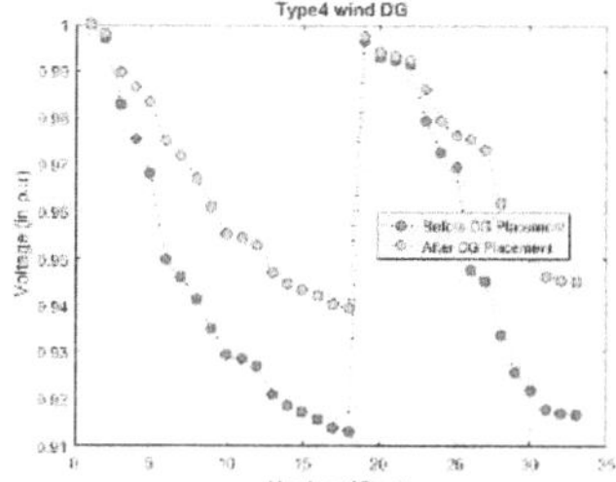
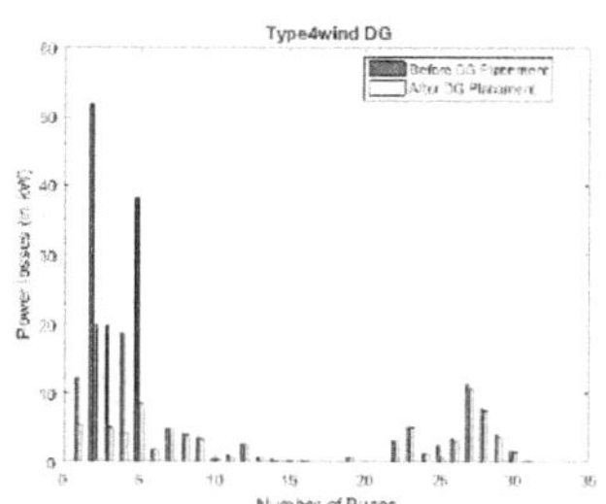

Figure 4.20 : Voltage profile and power losses of the test system for wind integration for scenario 4

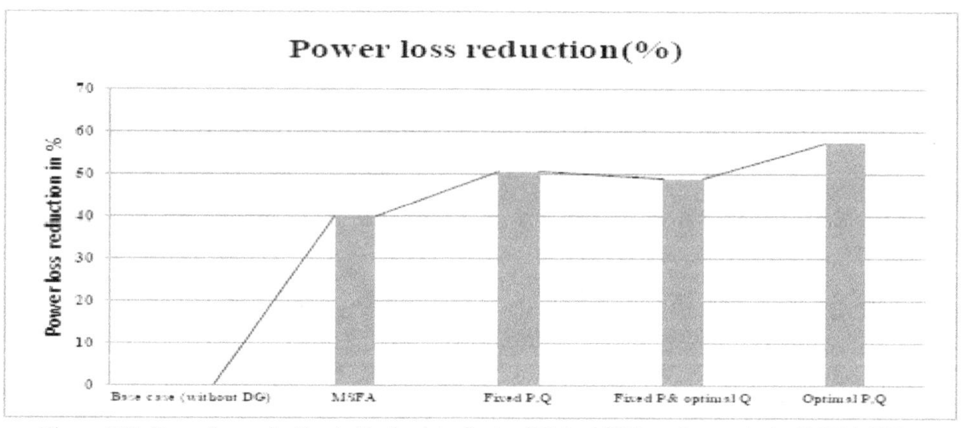

Figure 4.21: Power loss reduction in % after introducing DG for MSFA and scenario 4 of MSFA-PSO

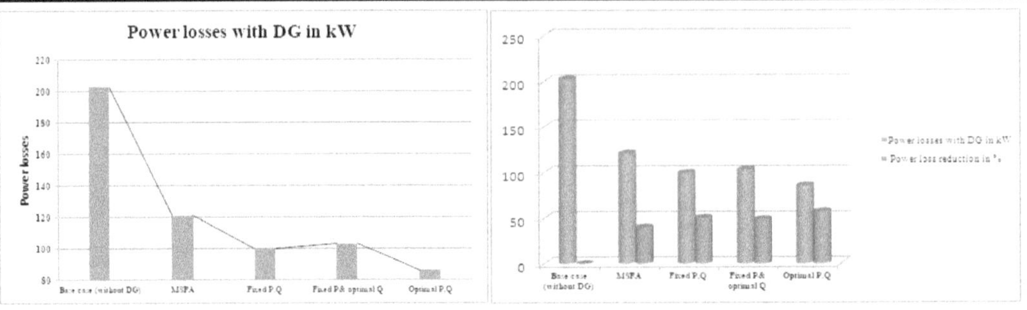

Figure: 4.22: Power loss reduction in kW & % after introducing DG for MSFA and scenario 4 of MSFA-PSO

Case-2, Scenario-5: 50 % integration of a wind DG

The results from developed MSFA-PSO algorithm for allocation of a wind DG unit for scenario-1 are tabulated in table below. the analysis is performed for three different cases which is further compared to the previously calculated value of the modified shuffled frog leap algorithm. The first technique allocates the DG according to the fixed value of active and reactive power of the introduced wind generator in the system. Here, the system losses are reduced to 98.01 kW from 202.68 kW and percentage losses are reduced to 51.63. the next technique allocates the DG according to the fixed value of active power and optimal value of the reactive power which is calculated by the algorithm. here, the system losses reduced are to 100.87 kW from 202.68 kW and percentage losses are reduced to 50.22. the last approach allocates the DG according to the optimal value of the active and reactive power which is calculated by the algorithm. here, the system losses reduced are to 76.66 kW from 202.68 kW and percentage losses are reduced to 62.17. Minimum voltage before DG placement is 0.91309 p. u at bus 18 and after DG placement the minimum voltage obtained from the best of the three approaches is observed to be 0.96132 p.u.at bus 18. Figures shows the graphs of voltage and power losses at different buses before and after integration of 50% of wind DG.

Table 4.13: Allocation of wind DG for scenario 5

Particulars	MSFA 50% wind DG	MSFA-PSO 50% wind DG (fixed P, Q)	MSFA-PSO 50% wind DG (Fixed P, Optimal Q within 50%)	MSFA-PSO wind DG (Optimal P , Q within 50%)
1: OLDG	Bus 12	Bus 33	Bus 08	Bus 31
2: ODGS	1857.5 kW	1857.5 kW	1857.5 kW	1586 kW

	1094.1 kVAR	-1094.1 kVAR	-834.87 kVAR	-940 kVAR
3: BCPL (kW)	202.68	202.68	202.68	202.68
4: PLDG (kW)	108.08	98.019	100.87	76.66
5: PLR (%)	46.67	51.63	50.22	62.17
6: Min V w/o DG (p.u.)	Bus 18	Bus 18	Bus 18	Bus 18
	0.91309	0.91309	0.91309	0.91309
7: Min V W DG (p.u.)	Bus 33	Bus 18	Bus 18	Bus 18
	094668	0.95931	0.95321	0.96132

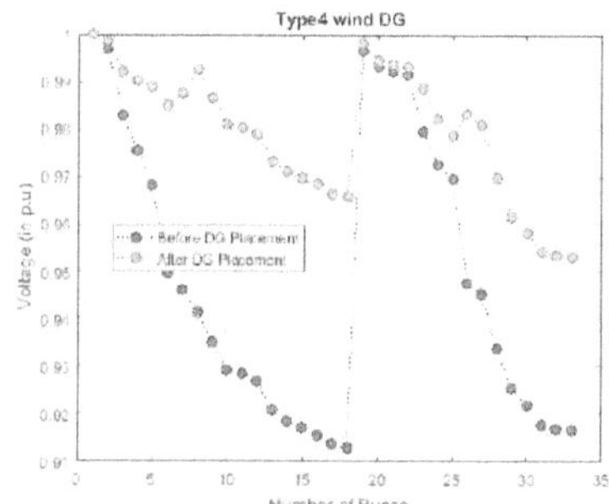

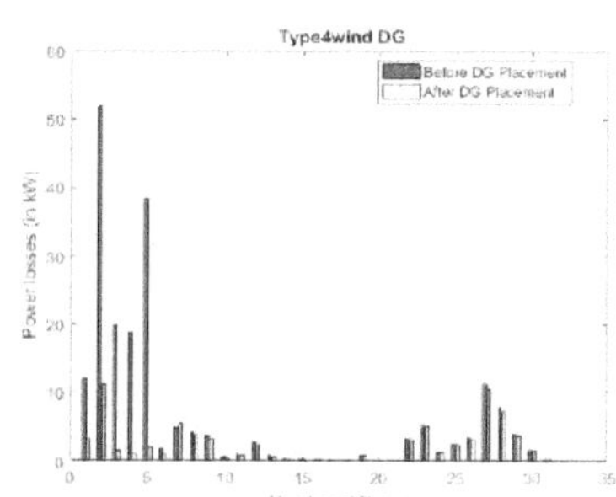

Figure 4.23 : Voltage profile and power losses of the test system for wind integration for scenario 5

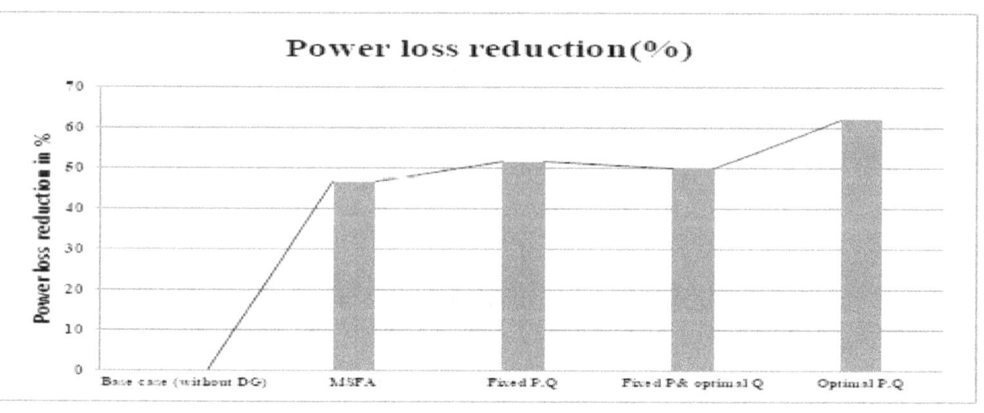

Figure: 4.24: Power loss reduction in % after introducing DG for MSFA and scenario 5 of MSFA-PSO

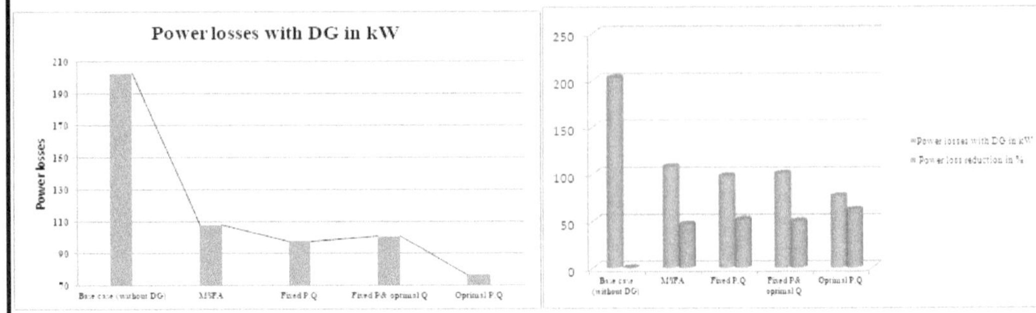

Figure 4.25: Power loss reduction in kW & % after introducing DG for MSFA and scenario 5 of MSFA-PSO

Case-2, Scenario 1- : Optimal placement of a WIND DG

The results obtained from the developed MSFA-PSO algorithm for case-2, scenario-1 and comparison with SFA is tabulated in table.4.3. Total losses of the system are reduced to 61.23 kW from 202.68 kW with percentage losses are reduced by 69.7%.

Table 4.14: Optimal allocation of Wind DG by MSFA-PSO and compared with MSFA

Particulars	MSFA WIND DG	MSFA-PSO WIND DG
1: OLDG	Bus 31	Bus 07
2: ODGS	1.3559 kW, -0.6136 kVAR	2226 kW, -1670.8 kVAR
3: BCPL (kW)	202.68	202.68
4: PLDG (kW)	63.82	61.23
5: PLR (%)	68.5	69.7
6: Min V w/o DG (p.u.)	Bus 18 0.91309	Bus 18 0.91309
7: Min V W DG (p.u.)	Bus 18 0.9398	Bus 33 0.96492

Case-2, Scenario-2: optimal placement of Two WIND DG units

The results obtained from the developed MSFA-PSO algorithm for case-2, scenario-2 is tabulated in table.4.16. Total losses of the system are reduced to 78.38 kW from 202.68 kW.

Table 4.15: Optimal allocation of two wind3 DG by MSFA-PSO

Particulars	MSFA-PSO - 2 WIND DG
1: OLDG	Bus 30, Bus 7
2: ODGS	977.01 kW, -1236.6 kVAR 1.2364 kW, -0.5176 kVAR
3: BCPL (kW)	202.68
4: PLDG (kW)	78.32
5: PLR (%)	61.35
6: Min V w/o DG (p.u.)	Bus 18 0.91309
7: Min V W DG (p.u.)	Bus 18 0.96636

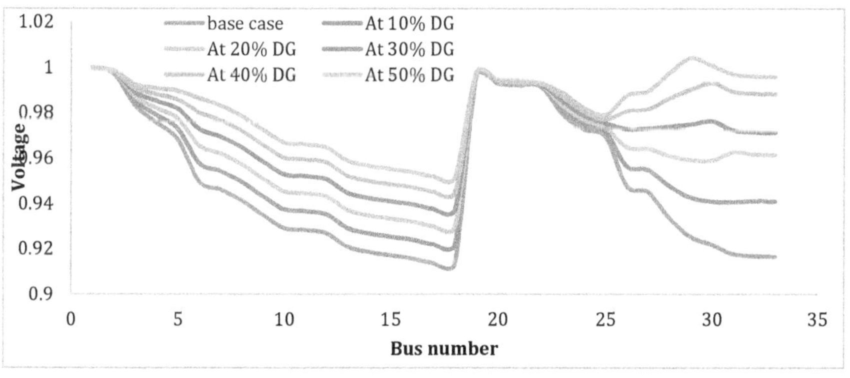

Figure 4.26: Voltage variation in the test system from 10% to 50% penetration of wind DG

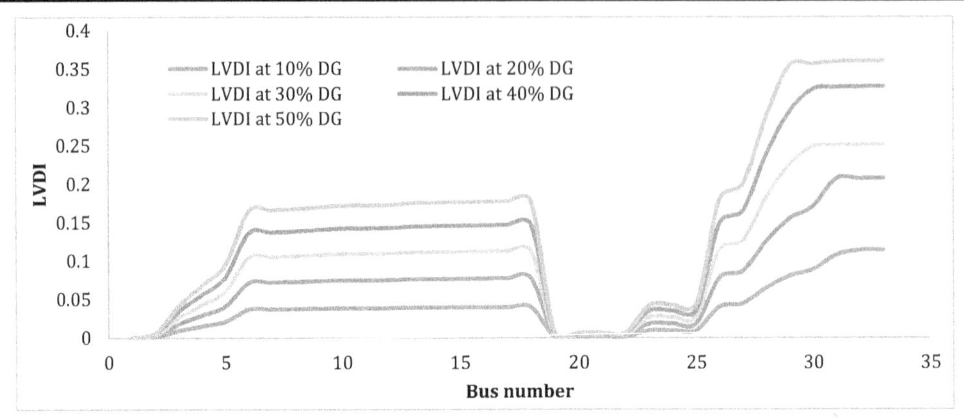

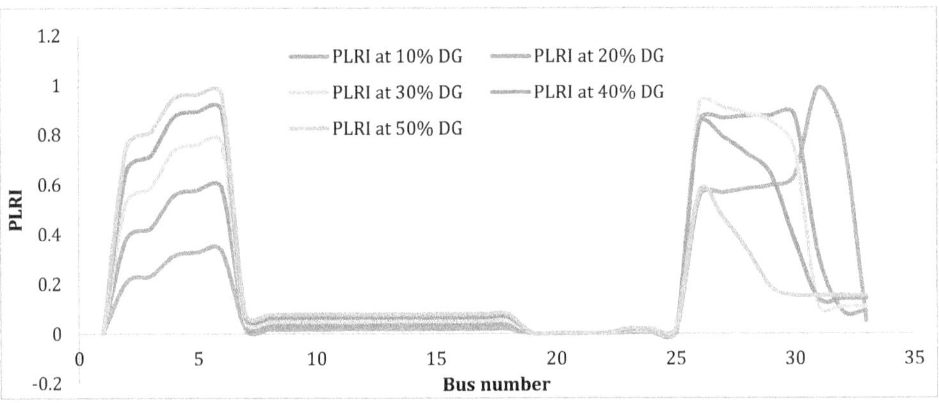

Figure 4.27: LVDI AND PLRI values in the test system from 10% to 50% penetration of wind DG

Table 4.16: WSOF values at different buses after integration of wind DG at different penetration levels

Bus number	WSOF at 10%	WSOF at 20%	WSOF at 30%	WSOF at 40%	WSOF at 50%
1	0	0	0	0	0
2	0.084081	0.155035	0.21492	0.265644	0.304627
3	0.09834	0.180463	0.248797	0.305914	0.348746
4	0.133945	0.239485	0.319915	0.381556	0.419307
5	0.143408	0.255634	0.340272	0.404109	0.442035
6	0.155669	0.278229	0.371327	0.442834	0.486257
7	0.029212	0.056651	0.082672	0.10787	0.130249

8	0.029505	0.057209	0.083484	0.108902	0.131459
9	0.02989	0.057923	0.084518	0.110246	0.13307
10	0.030216	0.058544	0.08538	0.111361	0.134427
11	0.030255	0.058642	0.08554	0.111566	0.134666
12	0.030361	0.058817	0.085792	0.111887	0.135044
13	0.030669	0.059421	0.086712	0.113056	0.13647
14	0.030796	0.059699	0.087069	0.113524	0.137019
15	0.030902	0.059852	0.087302	0.113851	0.137383
16	0.030985	0.060042	0.087554	0.114142	0.137753
17	0.031067	0.060229	0.087826	0.114522	0.138217
18	0.031119	0.060302	0.08794	0.114683	0.138382
19	0.001111	0.00217	0.003202	0.004184	0.005112
20	0.001104	0.002198	0.003225	0.004221	0.005114
21	0.001144	0.002196	0.003248	0.004247	0.005167
22	0.001124	0.002221	0.003256	0.004247	0.005166
23	0.007251	0.014124	0.020778	0.027123	0.032951
24	0.007346	0.014286	0.021	0.027414	0.033297
25	0.007379	0.014385	0.021126	0.027586	0.033508
26	0.244418	0.388652	0.43783	0.426237	0.336486
27	0.253414	0.398156	0.439337	0.415045	0.306107
28	0.271827	0.424739	0.463695	0.432036	0.307063
29	0.286554	0.443564	0.47568	0.432141	0.286433
30	0.306808	0.453713	0.445459	0.343949	0.274645
31	0.459608	0.24976	0.193913	0.251567	0.276676
32	0.392648	0.160799	0.194227	0.252002	0.277123
33	0.088594	0.160877	0.194319	0.252122	0.277256

Case-3, Scenario-1: 10 % integration of SOLAR, WIND Hybrid DG

The results from developed MSFA-PSO algorithm for allocation of Hybrid DG unit for scenario-1 are tabulated in table 4.17 The system losses reduced to 103.25 kW from 202.68 kW and percentage losses are reduced to 49.05%. Minimum voltage before DG placement is 0.91309 p.u at bus 18 and after DG placement the minimum voltage obtained to be 0.9472 p.u.at bus 31.Figures show the variation of voltage and power losses at different buses before and after integration of 10% of hybrid DG.

Table 4.17: Allocation of Hybrid DG for scenario 1

Particulars	MSFA Hybrid DG	MSFA-PSO Hybrid DG
1: OLDG	Bus 08 - Solar Bus 33 - Wind	Bus 12 – Solar Bus 33 - Wind
2: ODGS	743 kW 371.5 kW, -79.12 kVAR	743 kW 371.5 kW, -79.12 kVAR
3: BCPL (kW)	202.68	202.68
4: PLDG (kW)	139.23	103.25
5: PLR (%)	31.30	49.05
6: Min V w/o DG (p.u.)	Bus 18 0.91309	Bus 18 0.91309
7: Min V W DG (p.u.)	Bus 33 0.93291	Bus 31 0.9472

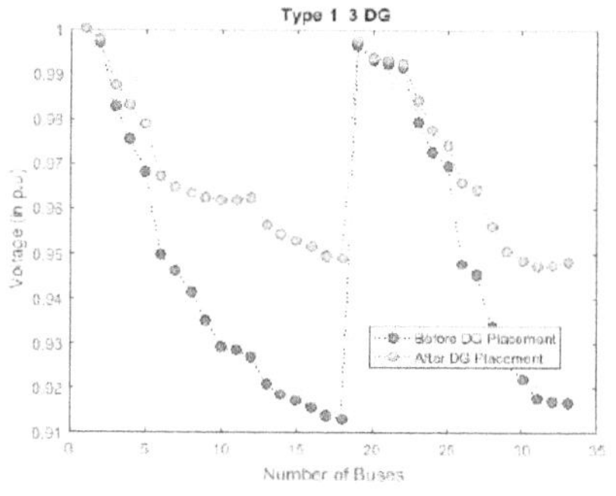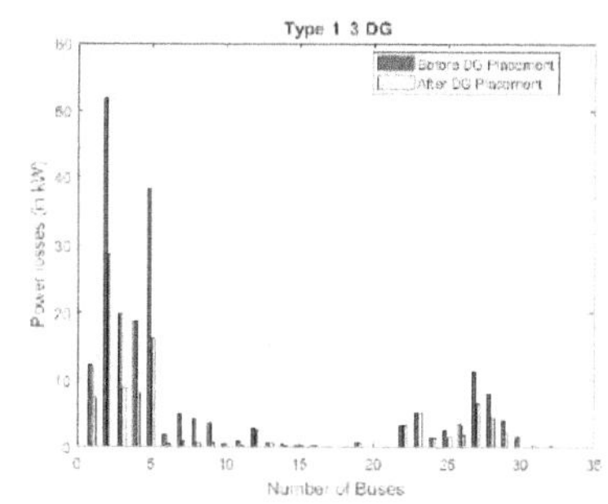

Figure 4.28 : Voltage profile and power losses of the IEEE33 bus system for wind and solar integration for scenario 1(10% penetration)

Case-3, Scenario-2: 20 % integration of SOLAR, WIND Hybrid DG

The results from developed MSFA-PSO algorithm for allocation of Hybrid DG unit for scenario-2 are tabulated in table 4.18 The system losses reduced to 73.63 kW from 202.68 kW and percentage losses are reduced to 63.66%. Minimum voltage before DG placement is 0.91309 p.u at bus 18 and after DG placement the minimum voltage obtained to be 0.98043 p.u.at bus 25.

Table 4.18: Allocation of Hybrid DG for scenario 2

Particulars	MSFA Hybrid DG	MSFA-PSO Hybrid DG
1: OLDG	Bus 08 - Solar Bus 33 - Wind	Bus 12 – Solar Bus 33 - Wind
2: ODGS	743 kW, 743 kW, -460 kVAR	743 kW, 743 kW, - 460 kVAR
3: BCPL (kW)	202.68	202.68
4: PLDG (kW)	129.32	73.63
5: PLR (%)	36.19	63.66
6: Min V w/o DG (p.u.)	Bus 18 0.91309	Bus 18 0.91309
7: Min V W DG (p.u.)	Bus 18 0.95961	Bus 25 0.98043

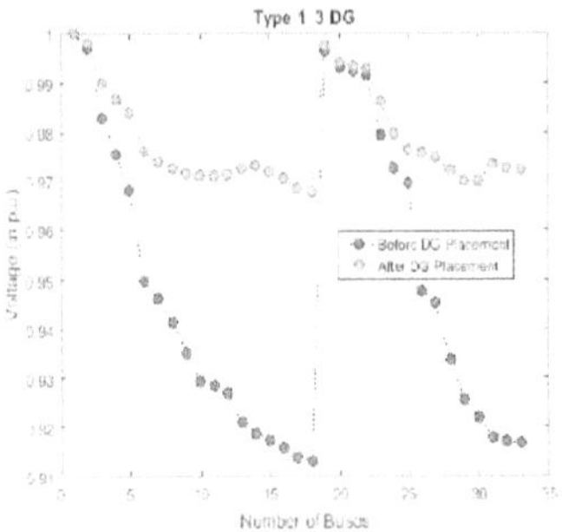

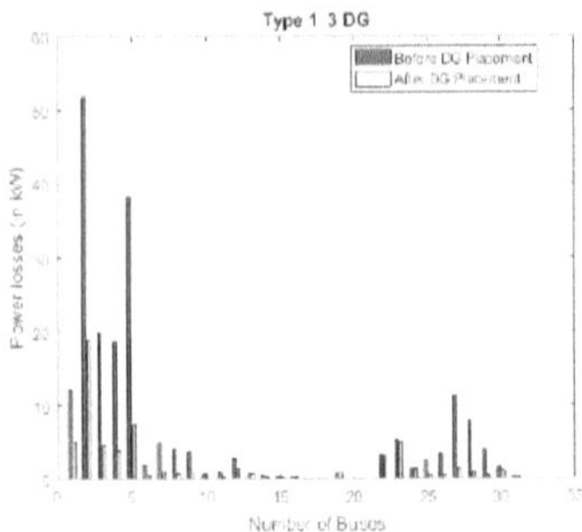

Figure 4.29 : Voltage profile and power losses of the IEEE33 bus system for wind and solar integration for scenario 2 (20% penetration)

Figure shows the variation of voltage and power losses at different buses before and after integration of 20% of hybrid DG.

Case-3, Scenario-3: 30 % integration of SOLAR, WIND Hybrid DG

The results from developed MSFA-PSO algorithm for allocation of Hybrid DG unit for scenario-3 are tabulated in table 4.19. The system losses reduced to 69.44 kW from 202.68 kW and percentage losses are reduced to 65.49%. Minimum voltage before DG placement is 0.91309 p.u at bus 18 and after DG placement the minimum voltage obtained to be 0.96292 p.u.at bus 18.Figures 4.30 shows the variation of voltage and power losses at different buses before and after integration of 30% of hybrid DG.

Table 4.19: Allocation of Hybrid DG for scenario 3

Particulars	MSFA Hybrid DG	MSFA-PSO Hybrid DG
1: OLDG	Bus 30 - Solar Bus 02 - Wind	Bus 26 - Solar Bus 29 - Wind
2: ODGS	1486 kW, 1486 kW, -649.79 kVAR	1486 kW, 1486 kW, -649.79 kVAR
3: BCPL (kW)	202.68	202.68
4: PLDG (kW)	105.94	69.94
5: PLR (%)	47.72	65.49
6: Min V w/o DG (p.u.)	Bus 18 0.91309	Bus 18 0.91309
7: Min V W DG (p.u.)	Bus 18	Bus 18

	0.941888	0.96292

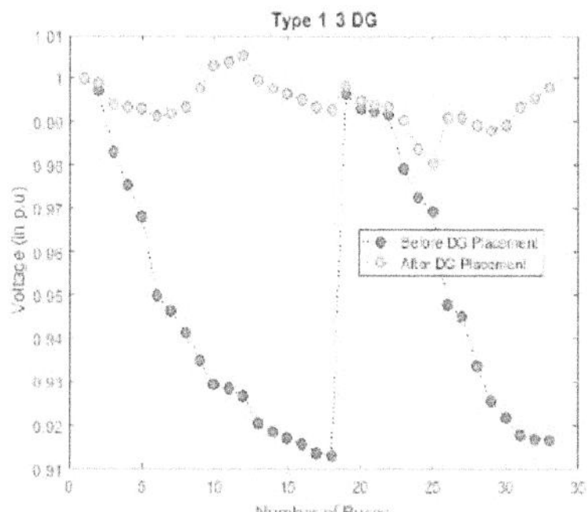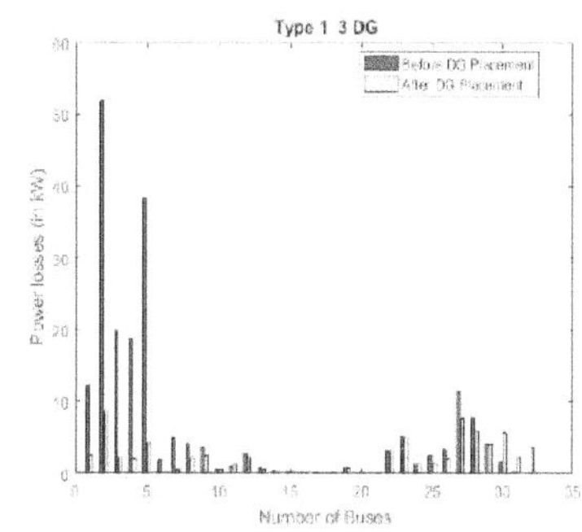

Figure 4.30 : Voltage profile and power losses
of the IEEE33 bus system for wind and solar integration for scenario 3 (30% penetration)

Case-3, Scenario-4: 40 % integration of SOLAR, WIND Hybrid DG

The results from developed MSFA-PSO algorithm for allocation of Hybrid DG unit for scenario-4 are tabulated in table 4.20. The system losses reduced to 66.07 kW from 202.68 kW and percentage losses are reduced to 67.39%. Minimum voltage before DG placement is 0.91309 p.u at bus 18 and after DG placement the minimum voltage obtained to be 0.95751 p.u.at bus 18. Figure shows the variation of voltage and power losses at different buses before and after integration of 40% of hybrid DG.

Table 4.20: Allocation of Hybrid DG for scenario 4

Particulars	MSFA Hybrid DG	MSFA-PSO Hybrid DG
1: OLDG	Bus 06 Bus 30	Bus 23 Bus 06
2: ODGS	1857.5 kW, 1857.5 kW, -1005.5 kVAR	1857.5 kW, 1857.5 kW, -1005.5 kVAR
3: BCPL (kW)	202.68	202.68
4: PLDG (kW)	95.23	66.078
5: PLR (%)	53.01	67.39
6: Min V w/o DG (p.u.)	Bus 18	Bus 18

	0.91309	0.91309
7: Min V W DG (p.u.)	Bus 18	Bus 18
	0.942951	0.95751

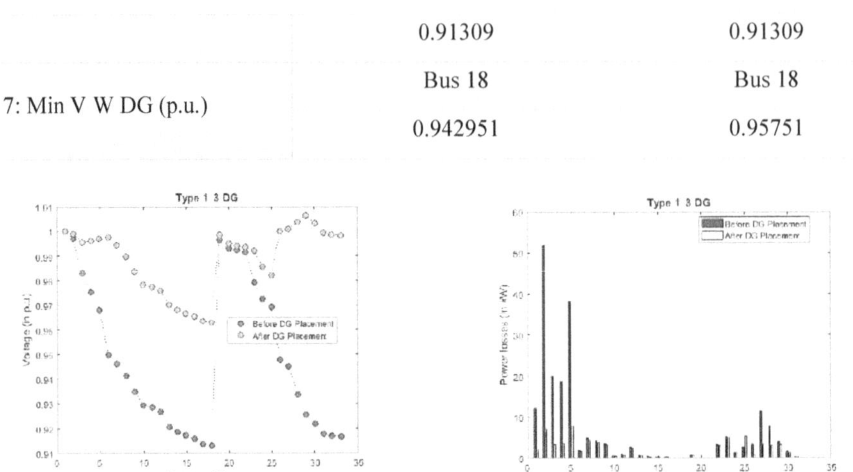

Figure 4.31 : Voltage profile and power losses of the IEEE33 bus system for wind and solar integration for scenario 4 (40% penetration)

Case-3, Scenario-5: 50 % integration of SOLAR, WIND Hybrid DG

The results from developed MSFA-PSO algorithm for allocation of Hybrid DG unit for scenario-5 are tabulated in table 4.21. The system losses reduced to 60.86 kW from 202.68 kW and percentage losses are reduced to 69.96%. Minimum voltage before DG placement is 0.91309 p.u at bus 18 and after DG placement the minimum voltage obtained to be 0.96791 p.u.at bus 18.Figure shows the variation of voltage and power losses at different buses before and after integration of 50% of hybrid DG.

Table 4.21: Allocation of Hybrid DG for scenario 5

Particulars	MSFA Hybrid DG	MSFA-PSO Hybrid DG
1: OLDG	Bus 02 Bus 30	Bus 23 Bus 06
2: ODGS	1857.5 kW, 1857.5 kW, -1005 kVAR	1857.5 kW, 1857.5 kW, -1005 kVAR
3: BCPL (kW)	202.68	202.68
4: PLDG (kW)	89.39	60.86
5: PLR (%)	55.90	69.96
6: Min V w/o DG (p.u.)	Bus 18	Bus 18

	0.91309	0.91309
7: Min V W DG (p.u.)	Bus 18	Bus 18
	0.9569	0.96791

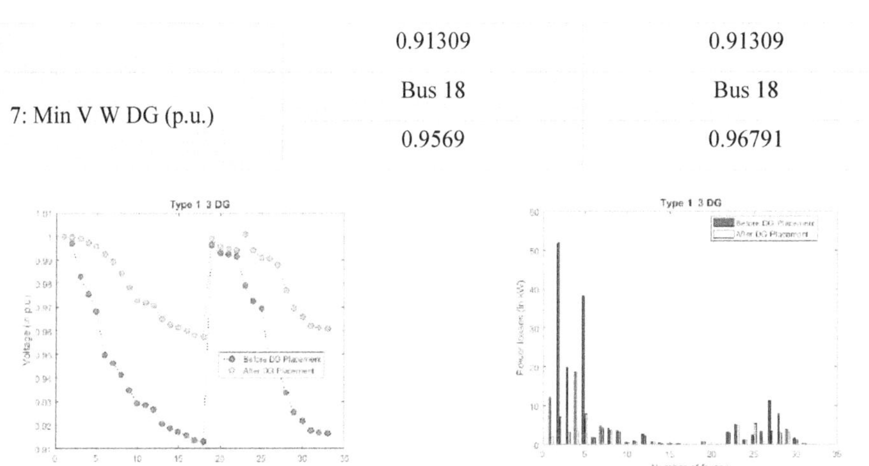

Figure 4.32 : Voltage profile and power losses of the IEEE33 bus system for wind and solar integration for scenario 5 (50% penetration)

Case-3: Optimal placement of Solar and Wind (Hybrid) DG units

The results obtained from the developed MSFA-PSO algorithm for case-3 is tabulated in table.4.22. Total losses of the system are reduced to 66.23 kW from 202.68 kW.

Table 4.22: Optimal allocation of Solar and wind hybrid DG by MSFA-PSO

Particulars	MSFA-PSO Hybrid DG
1: OLDG	Bus 26 Bus 32
2: ODGS	1939 kW 300 kW, -111.79 kVAR
3: BCPL (kW)	202.68
4: PLDG (kW)	66.23
5: PLR (%)	67.32
6: Min V w/o DG (p.u.)	Bus 18 0.91309
7: Min V W DG (p.u.)	Bus 18 0.96636

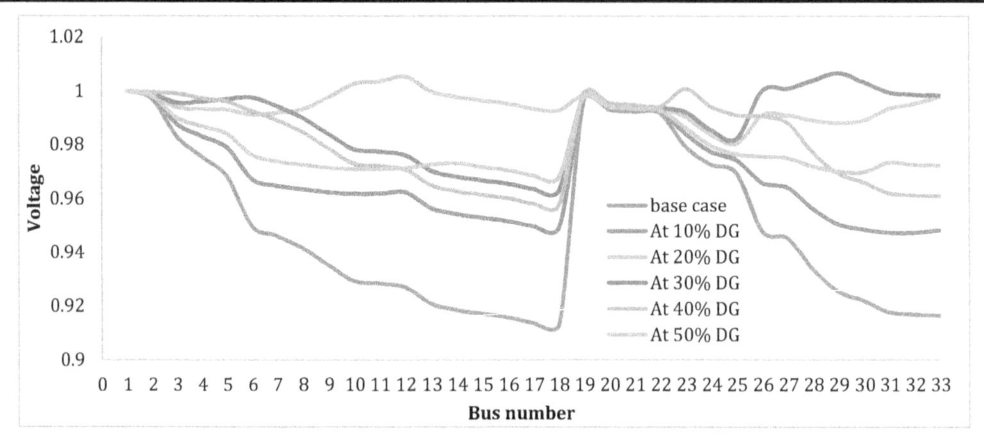

Figure: 4.33: Voltage variation in the test system from 10% to 50% penetration for Solar and wind hybrid DG

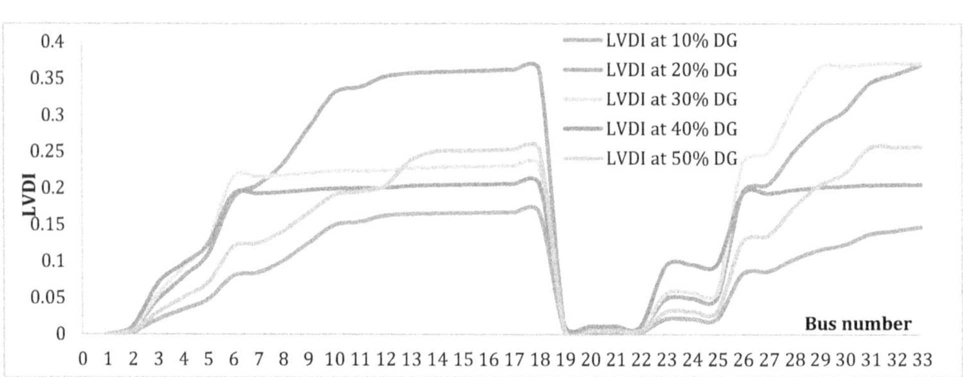

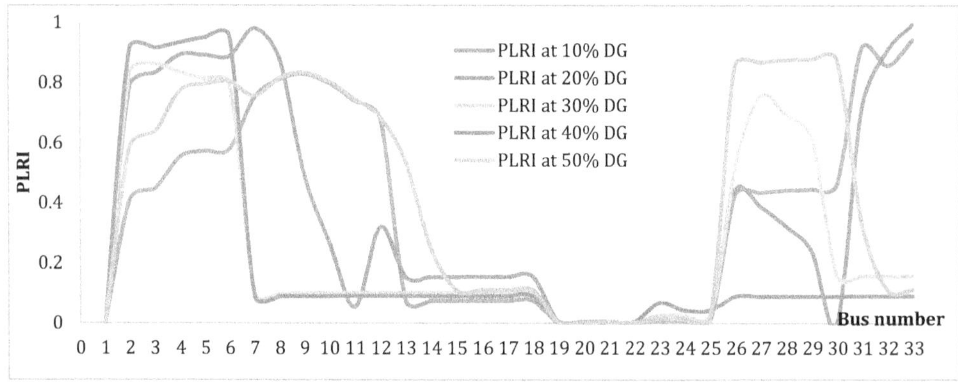

Figure: 4.34: LVDI & PLRI values in the test system from 10% to 50% penetration for Solar and wind hybrid DG

Table 4.23: WSOF values at different buses after integration of wind DG at different penetration levels

Bus No	WSOF at 10%	WSOF at 20%	WSOF at 30%	WSOF at 40%	WSOF at 50%
1	0	0	0	0	0
2	0.165986	0.31955	0.338382	0.372509	0.237821
3	0.19148	0.361674	0.378484	0.408539	0.273646
4	0.24166	0.404037	0.388336	0.431397	0.339908
5	0.257353	0.422813	0.401704	0.454032	0.360587
6	0.279285	0.466493	0.447092	0.49724	0.39358
7	0.350891	0.515092	0.16836	0.150648	0.377641
8	0.384947	0.492586	0.169933	0.152089	0.411333
9	0.406745	0.367594	0.172006	0.153938	0.432909
10	0.409301	0.306591	0.173715	0.155484	0.43566
11	0.388298	0.224721	0.174017	0.155764	0.414954
12	0.369537	0.339553	0.174516	0.156206	0.396433
13	0.128287	0.275488	0.176303	0.157831	0.357153
14	0.12882	0.276543	0.177018	0.158442	0.250316
15	0.129165	0.277271	0.177499	0.15889	0.194586
16	0.129534	0.277954	0.177955	0.159301	0.195083
17	0.129975	0.278902	0.178559	0.159844	0.19575
18	0.130116	0.279209	0.178786	0.160033	0.195959
19	0.002557	0.005887	0.006841	0.008671	0.003745
20	0.002542	0.005904	0.006889	0.008705	0.003774
21	0.002564	0.005929	0.006901	0.008753	0.003827
22	0.002566	0.005927	0.006916	0.008737	0.003805
23	0.01644	0.037938	0.044104	0.083101	0.024304
24	0.016617	0.038368	0.044562	0.074277	0.024573
25	0.016706	0.038599	0.044834	0.074726	0.024715
26	0.220322	0.293772	0.346944	0.150734	0.418988
27	0.225785	0.279063	0.451474	0.151411	0.428443
28	0.238486	0.278944	0.466584	0.154287	0.455155
29	0.247517	0.265439	0.465863	0.156364	0.47404
30	0.257802	0.183403	0.282519	0.157294	0.484205

31	0.448844	0.49267	0.284599	0.158507	0.285015
32	0.428248	0.576879	0.285078	0.158804	0.198101
33	0.465495	0.619664	0.285221	0.158907	0.198199

Table 4.24: Comparison table for allocation of DG using different techniques

Method	P loss with DG (kW)	Loss reduction in %	Minimum voltage in p.u (Bus)	(bus) DG location	Optimal DG size (MW)	Optimal size (MVA)
GA [71]	107.1	49.71	0.9810 (25)	11	1.5	2.9942
				29	0.4228	
				30	1.0714	
PSO [71]	105.35	50.06	0.9806(30)	13	0.9816	2.9881
				32	0.8297	
				8	1.1768	
GA/PSO [71]	103.4	50.99	0.9808(25)	32	1.2	2.988
				16	0.863	
				11	0.925	
SA [72]	82.03	59.12	0.9676(14)	6	1.1124	2.4677
				18	0.4874	
				30	0.8679	
BFOA [73]	89.9	57.38	0.9705(29)	14	0.6521	1.9176
				18	0.1984	
				32	1.0672	
IWO [102]	85.86	57.47	0.9716(29)	14	0.6247	1.7856
				18	0.1049	
				32	1.056	
MSFA	79.625	60.82	0.96517(14)	17	0.354.55	2.16552
				7	0.876.81	
				31	0.934.16	
MSFA- PSO	77.45	61.78	0.970351(18)	06	1.2535	2.01565
				18	0.9681	
				32	0.8981	

CASE 2: IEEE 69 BUS TEST SYSTEM

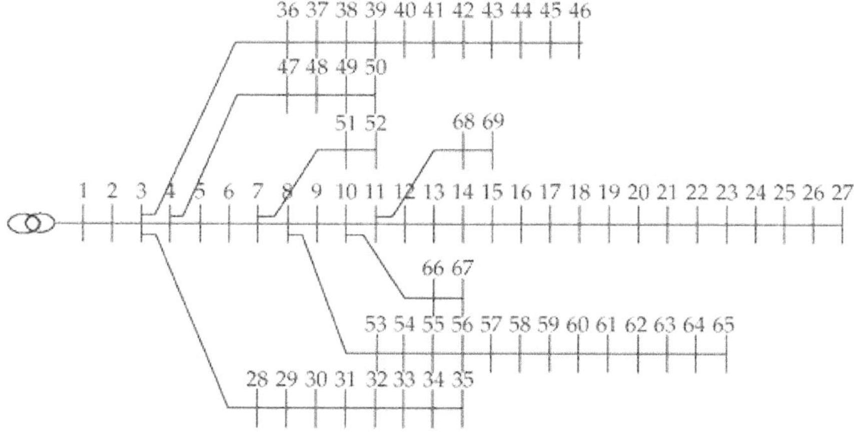

Figure 4.35: Single line diagram of the IEEE 69 bus test system

The developed MSFA-PSO method is applied to the standard IEEE 69 bus system operated at 12.66kV level of voltage and consists of 69 busses, 68 branches. And 3800kW and 2690kVar of total real and reactive power of the system respectively. Fig 4.35 shows the diagram of the network considered. The basic data of the network that is the line data and the branch data are given in appendix B. The nodes with the highest amount of deviation are identified and tabulated in the table. The base case load flow is obtained where the power losses without the connection of the DG's are obtained in kW and kVAR.

The following cases and scenarios are considered for analysis of the system

Case -1: Allocation of Solar DG unit

Case -2: Allocation of Wind DG unit

Case -3: Allocation of Hybrid DG unit

Scenario 1: 10 % Penetration of DG

Scenario 2: 20 % Penetration of DG

Scenario 3: 30 % Penetration of DG

Scenario 4: 40 % Penetration of DG

Scenario 5: 50 % Penetration of DG

Case -4: Optimal Allocation of Solar DG unit

Case -5: Optimal Allocation of Wind DG unit

Scenario 1: Placement of single DG

Scenario 2: Placement of Two DGs

Case -6: Optimal Allocation of Hybrid DG unit

Case-1, Scenario-1: 10 % integration of a SOLAR DG

The results from developed MSFA-PSO algorithm for allocation of a Solar DG unit for scenario-1 are tabulated in table 4.26. The system losses reduced to 169.77 kW from 224.98 kW and percentage losses are reduced by 24.54%. Minimum voltage before DG placement is 0.90918 p. u at bus 65 and after DG placement it the minimum voltage obtained to be 0.92677 p.u.at bus 65.

Table 4.26: Allocation of Solar DG for scenario 1

Particulars	MSFA 10% SOLAR DG	MSFA-PSO 10% SOLAR DG
1: OLDG	Bus 62	Bus 61
2: ODGS	380 kW	380 kW
3: BCPL (kW)	224.98	224.98
4: PLDG (kW)	172.31	169.77
5: PLR (%)	23.41	24.54
6: Min V w/o DG (p.u.)	Bus 65	Bus 65
	0.90918	0.90918
7: Min V W DG (p.u.)	Bus 65	Bus 65
	0.92231	0.92677

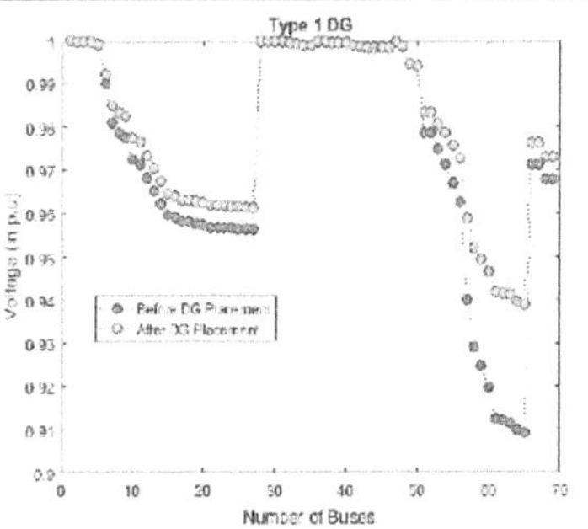
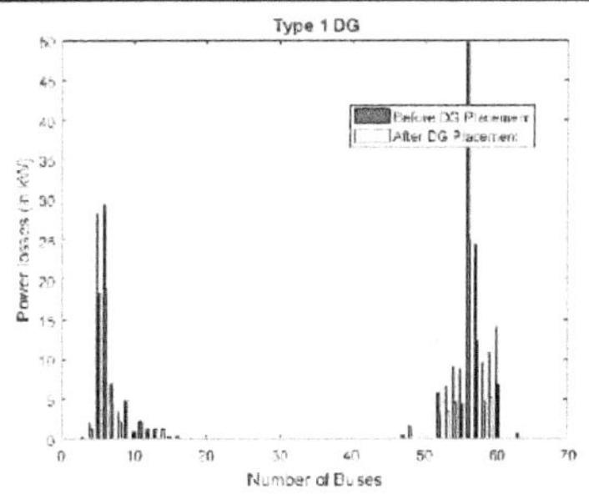

Figure 4.36: Voltage profile and power loss of the IEEE 69 bus test system of Solar DG for scenario 1

Case-1, Scenario-2: 20 % integration of a SOLAR DG

The results from developed MSFA-PSO algorithm for allocation of a Solar DG unit for scenario-2 are tabulated in table 4.30. The system losses reduced to 130.17 kW from 224.98 kW and percentage losses are reduced by 42.14%. Minimum voltage before DG placement is 0.90918 p. u at bus 65 and after DG placement it the minimum voltage obtained to be 0. 93877 p.u.at bus 65.

Table 4.27: Allocation of Solar DG for scenario 2

Particulars	MSFA 20% SOLAR DG	MSFA-PSO 20% SOLAR DG
1: OLDG	Bus 62	Bus 61
2: ODGS	760 kW	760 kW
3: BCPL (kW)	224.98	224.98
4: PLDG (kW)	132.51	130.17
5: PLR (%)	41.10	42.14
6: Min V w/o DG (p.u.)	Bus 65	Bus 65
	0.90918	0.90918
7: Min V W DG (p.u.)	Bus 65	Bus 65
	0.93456	0.93887

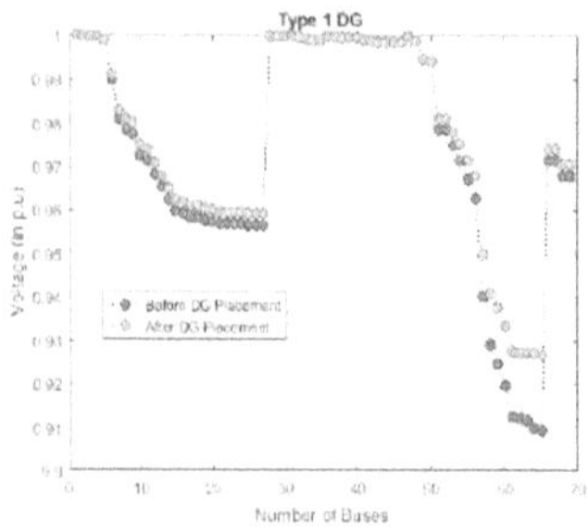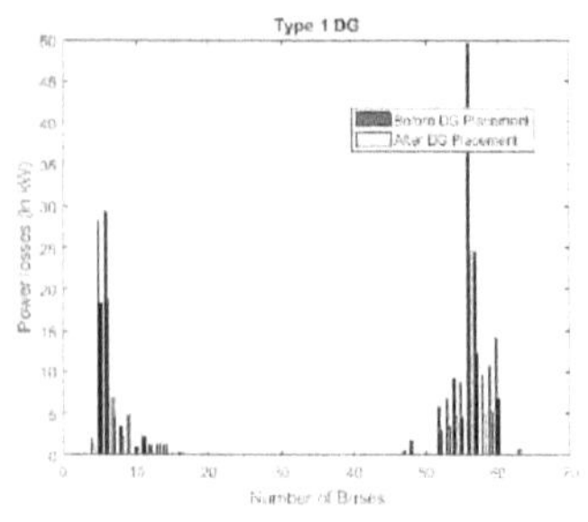

Figure 4.37: Voltage profile and power loss of the IEEE 69 bus test system of Solar DG for scenario 2

Case-1, Scenario-3: 30 % integration of a SOLAR DG

The results from developed MSFA-PSO algorithm for allocation of a Solar DG unit for scenario-3 are tabulated in table 4.28 The system losses reduced to 102.97 kW from 224.98 kW and percentage losses are reduced by 54.23%. Minimum voltage before DG placement is 0.90918 p. u at bus 65 and after DG placement it the minimum voltage obtained to be 0.95296 p.u.at bus 65.

Table 4.28: Allocation of Solar DG for scenario 3

Particulars	MSFA 30% SOLAR DG	MSFA-PSO 30% SOLAR DG
1: OLDG	Bus 61	Bus 61
2: ODGS	1140 kW	1140 kW
3: BCPL (kW)	224.98	224.98
4: PLDG (kW)	105.61	102.97
5: PLR (%)	53.05	54.23
6: Min V w/o DG (p.u.)	Bus 65	Bus 65
	0.90918	0.90918
7: Min V W DG (p.u.)	Bus 65	Bus 65
	0.94325	0.95296

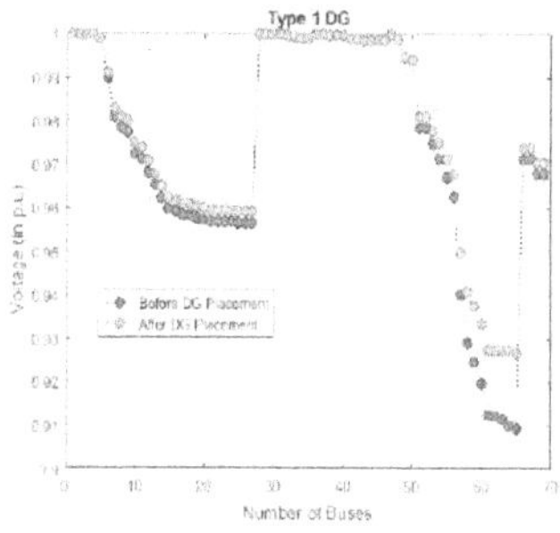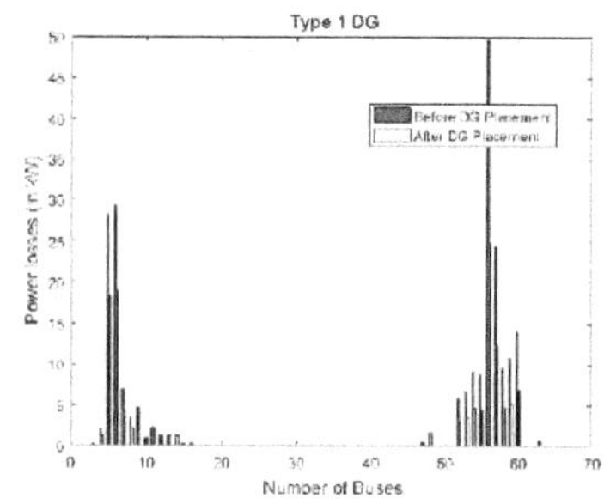

Figure 4.38: Voltage profile and power loss of the IEEE 69 bus test system of Solar DG for scenario 3

Case-1, Scenario-4: 40 % integration of a SOLAR DG

The results from developed MSFA-PSO algorithm for allocation of a Solar DG unit for scenario-4 are tabulated in table 4.29. The system losses reduced to 87.65 kW from 224.98 kW and percentage losses are reduced by 61.04%. Minimum voltage before DG placement is 0.90918 p.u at bus 65 and after DG placement it the minimum voltage obtained to be 0.96623p.u.at bus 27. Figure shows the variation of voltage and power losses at different buses before and after integration of 40% of Solar DG.

Table 4.29: Allocation of Solar DG for scenario 4

Particulars	MSFA 40% SOLAR DG	MSFA-PSO 40% SOLAR DG
1: OLDG	Bus 63	Bus 61
2: ODGS	1520 kW	1520 kW
3: BCPL (kW)	224.98	224.98
4: PLDG (kW)	90.13	87.65
5: PLR (%)	59.93	61.04
6: Min V w/o DG (p.u.)	Bus 65	Bus 65
	0.90918	0.90918
7: Min V W DG (p.u.)	Bus 27	Bus 27
	0.96234	0.96623

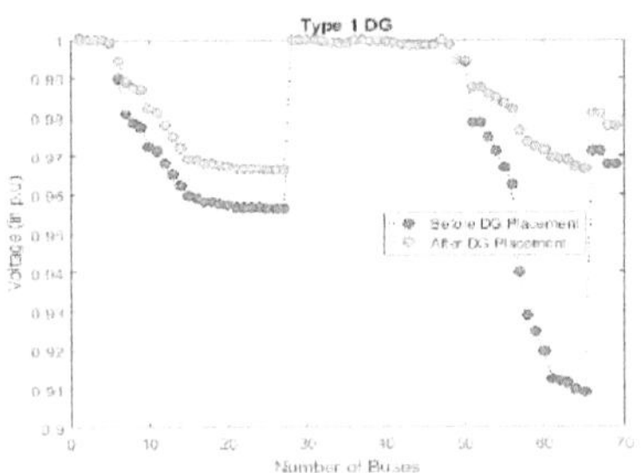
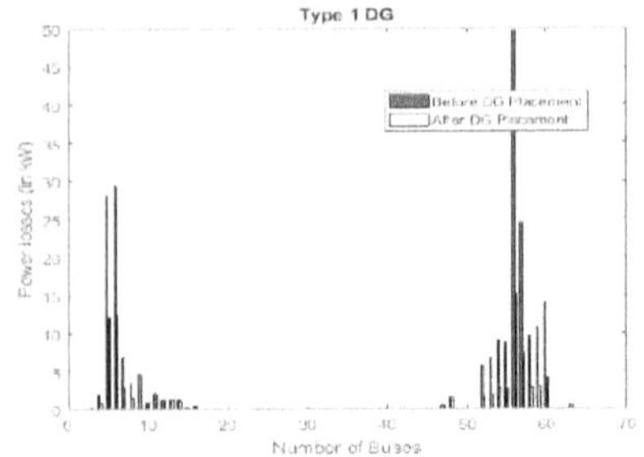

Figure 4.39: Voltage profile and power loss of the IEEE 69 bus test system of Solar DG for scenario 4

Case-1, Scenario-5: 50 % integration of a SOLAR DG

The results from developed MSFA-PSO algorithm for allocation of a Solar DG unit for scenario-5 are tabulated in table 4.30. The system losses reduced to 83.226 kW from 224.98 kW and percentage losses are reduced by 63.01%. Minimum voltage before DG placement is 0.90918 p. u at bus 65 and after DG placement it the minimum voltage obtained to be 0.96849 p.u.at bus 27. Figure 4.40 shows the variation of voltage and power losses at different buses before and after integration of 50% of Solar DG.

Table 4.30: Allocation of Solar DG for scenario 5

Particulars	MSFA 50% SOLAR DG	MSFA-PSO 50% SOLAR DG
1: OLDG	Bus 63	Bus 61
2: ODGS	1900 kW	1900 kW
3: BCPL (kW)	224.98	224.98
4: PLDG (kW)	84.31	83.226
5: PLR (%)	62.52	63.01
6: Min V w/o DG (p.u.)	Bus 65	Bus 65
	0.90918	0.90918
7: Min V W DG (p.u.)	Bus 27	Bus 27
	0.96521	0.96849

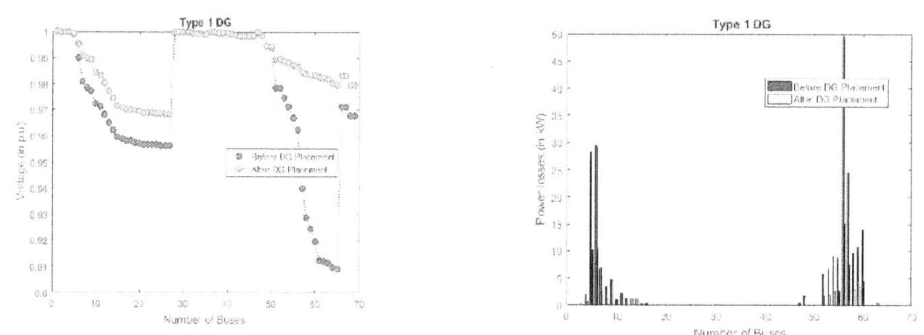

Figure 4.40: Voltage profile and power loss of the IEEE 69 bus test system of Solar DG for scenario 5

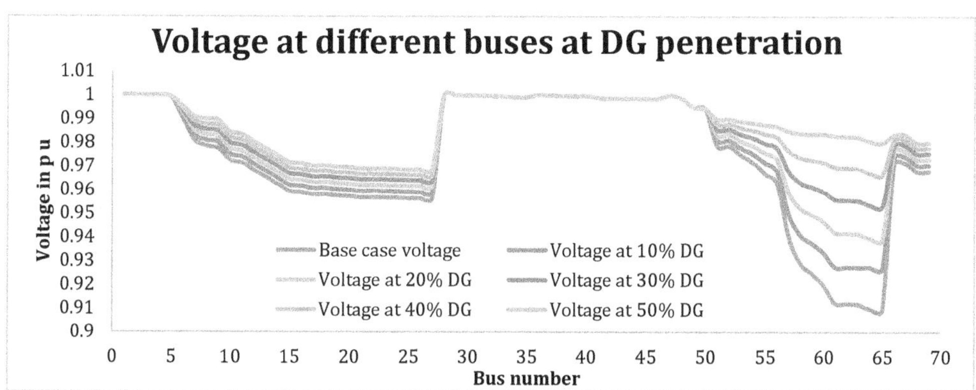

Figure 4.41: Voltage profile of the IEEE 69 bus test system for DG penetration from 10% to 50%

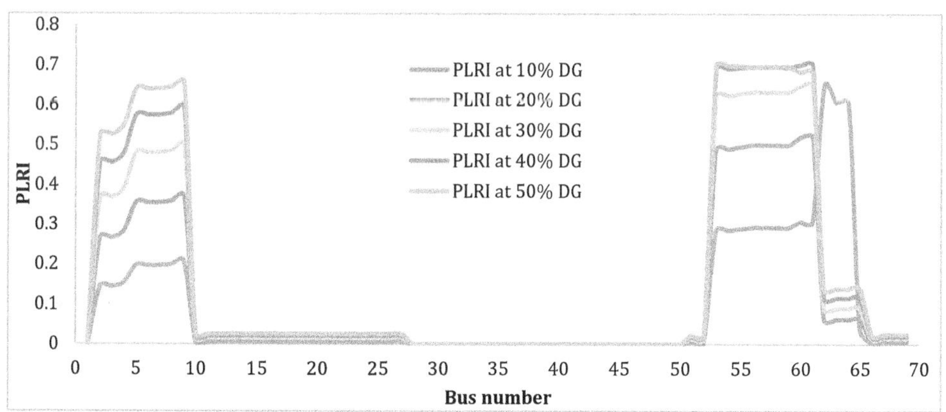

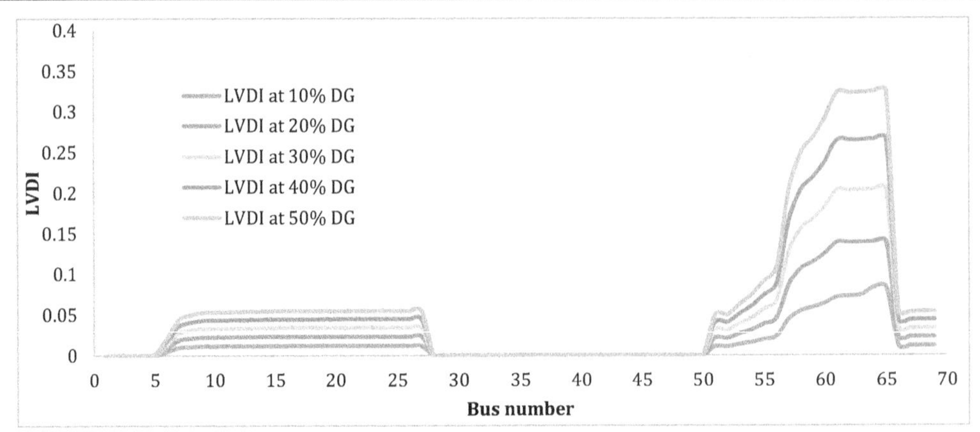

Figure 4.42: PLRI & LVDI values of the IEEE 69 bus test system for DG penetration from 10% to 50%

Table 4.31: WSOF values at different buses for integration of solar DG at different penetration level

Bus number	WSOF at 10%	WSOF at 20%	WSOF at 30%	WSOF at 40%	WSOF at 50%
1	0	0	0	0	0
2	0.057389	0.10618	0.147498	0.181935	0.210019
3	0.057389	0.10618	0.147498	0.181909	0.210019
4	0.061413	0.113116	0.156392	0.191871	0.220152
5	0.078717	0.142254	0.1924	0.230261	0.256838
6	0.081395	0.147485	0.200135	0.240383	0.269261
7	0.08433	0.153221	0.208502	0.251275	0.282605
8	0.086091	0.156333	0.212594	0.255982	0.287614
9	0.089751	0.162323	0.21984	0.263465	0.294337
10	0.009023	0.01766	0.025949	0.033927	0.041639
11	0.009042	0.017694	0.026022	0.034015	0.041731
12	0.009117	0.017812	0.026176	0.034247	0.04199
13	0.009189	0.017913	0.026343	0.03445	0.042236
14	0.009241	0.018012	0.026486	0.034606	0.04246
15	0.009237	0.018058	0.026552	0.034742	0.04261
16	0.009289	0.018128	0.026644	0.034812	0.042679

17	0.009301	0.018145	0.026656	0.034873	0.042745
18	0.009297	0.018155	0.026663	0.03486	0.042765
19	0.009271	0.01812	0.02666	0.034891	0.042776
20	0.009303	0.018181	0.026708	0.034899	0.04281
21	0.009322	0.018178	0.026736	0.03495	0.04286
22	0.009305	0.018187	0.02672	0.034938	0.042851
23	0.009305	0.018192	0.026724	0.034939	0.042853
24	0.009308	0.018192	0.02677	0.034955	0.042891
25	0.009311	0.018191	0.026732	0.034954	0.042866
26	0.009309	0.018204	0.026726	0.034952	0.042867
27	0.009313	0.018206	0.026735	0.034963	0.04288
28	0	0	1.15E-05	3.76E-05	3.76E-05
29	2.61E-05	2.61E-05	2.61E-05	5.21E-05	6.76E-05
30	2.61E-05	3.29E-05	3.29E-05	3.29E-05	6.59E-05
31	0	2.61E-05	2.61E-05	2.61E-05	5.21E-05
32	2.61E-05	3.38E-05	3.38E-05	5.99E-05	5.99E-05
33	3.25E-05	3.25E-05	5.86E-05	5.86E-05	5.86E-05
34	2.61E-05	2.61E-05	2.61E-05	2.61E-05	5.22E-05
35	0	0	0	3.44E-05	3.44E-05
36	0	2.61E-05	2.61E-05	2.61E-05	2.61E-05
37	0	0	2.61E-05	2.61E-05	2.61E-05
38	0	0	2.61E-05	2.61E-05	2.61E-05
39	3.41E-05	3.41E-05	3.41E-05	4.21E-05	6.81E-05
40	0	2.61E-05	2.61E-05	2.61E-05	4.62E-05
41	2.61E-05	2.61E-05	3.43E-05	3.43E-05	6.04E-05
42	0	2.61E-05	2.61E-05	2.61E-05	2.61E-05
43	2.61E-05	2.61E-05	2.61E-05	2.61E-05	6.72E-05
44	2.61E-05	2.61E-05	3.39E-05	3.39E-05	6E-05
45	6.58E-06	6.58E-06	6.58E-06	3.93E-05	3.93E-05
46	2.61E-05	2.61E-05	2.61E-05	5.22E-05	5.22E-05
47	2.61E-05	4.32E-05	6.93E-05	9.54E-05	0.000113
48	2.61E-05	5.91E-05	9.2E-05	9.89E-05	0.000132

49	2.62E-05	2.62E-05	7.69E-05	0.000103	0.000103
50	2.62E-05	5.24E-05	5.24E-05	0.000113	0.000139
51	0.008443	0.016465	0.024215	0.031654	0.038823
52	0.008436	0.016472	0.024201	0.031658	0.038828
53	0.121882	0.211472	0.272383	0.306933	0.317319
54	0.123688	0.214856	0.277147	0.312927	0.324367
55	0.127307	0.221137	0.285222	0.322019	0.333764
56	0.13081	0.227226	0.293092	0.330907	0.342983
57	0.143273	0.25146	0.328574	0.377132	0.399488
58	0.149611	0.263756	0.346542	0.400489	0.427997
59	0.152116	0.268632	0.353616	0.409682	0.439179
60	0.161095	0.282619	0.369014	0.42319	0.447754
61	0.165405	0.290928	0.381146	0.438847	0.466821
62	0.303728	0.108103	0.157689	0.204692	0.249422
63	0.28723	0.108185	0.15779	0.204847	0.249592
64	0.292525	0.108534	0.158296	0.205473	0.250368
65	0.064975	0.108635	0.158465	0.205697	0.250612
66	0.009031	0.017643	0.025923	0.033911	0.041569
67	0.009026	0.017637	0.025912	0.033908	0.041573
68	0.00909	0.017781	0.026139	0.034155	0.041876
69	0.00909	0.017769	0.026131	0.034152	0.04188

Case-2, Scenario-1: 10 % integration of a wind DG

The results from developed MSFA-PSO algorithm for allocation of a wind DG unit for scenario-1 are tabulated in table below. The analysis is performed for three different cases which is further compared to the previously calculated value of the modified shuffled frog leap algorithm. The first technique allocates the DG according to the fixed value of active and reactive power of the introduced wind generator in the system. Here, the system losses are reduced to 159 kW from 224.98 kW and percentage losses are reduced to 36.09. The next technique allocates the DG according to the fixed value of active power and optimal value of the reactive power which is calculated by the algorithm. Here, the system losses reduced are to 151.34 kW from 224.98 kW and percentage losses are reduced to 32.73.

The last approach allocates the DG according to the optimal value of the active and reactive power which is calculated by the algorithm. Here, the system losses reduced are to 157 kW from 224.98 kW and percentage losses are reduced to 30.21. Minimum voltage before DG placement is 0.90918 p. u at bus 65 and after DG placement the minimum voltage obtained from the best of the three approaches is observed to be 0.92725 p.u.at bus 65. Figure show the graphs of voltage and power losses at different buses before and after integration of 10% of wind DG.

Table 4.32: Allocation of wind DG for scenario 1

Particulars	MSFA 10% wind DG	MSFA-PSO 10% wind DG (fixed P, Q)	MSFA-PSO 10% wind DG (Fixed P, Optimal Q within 10%)	MSFA-PSO wind DG (Optimal P, Q within 10%)
1: OLDG	Bus 63	Bus 64	Bus 64	Bus 64
2: ODGS	380 kW, -269 kVAR	380 kW, -269 kVAR	380 kW, 129.36 kVAR	362.64 kW, -180.13 kVAR
3: BCPL (kW)	224.98	224.98	224.98	224.98
4: PLDG (kW)	145.36	159	151.34	157
5: PLR (%)	35.38	36.09	32.73	30.218
6: Min V w/o DG (p.u.)	Bus 65	Bus 65	Bus 65	Bus 65
	0.90918	0.90918	0.90918	0.90918
7: Min V W DG (p.u.)	Bus 65	Bus 65	Bus 65	Bus 65
	0.92221	0.92629	0.92725	0.92577

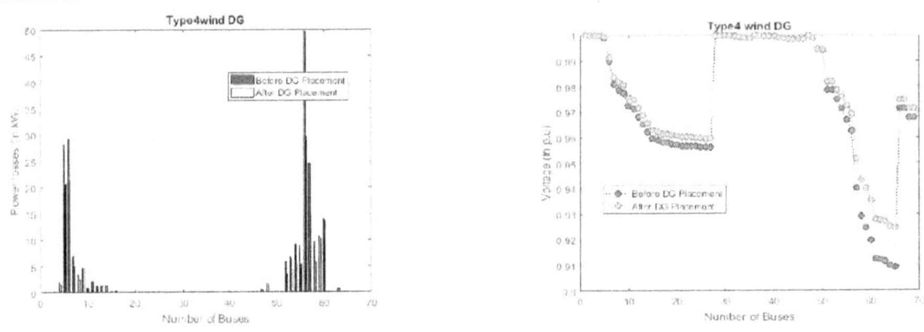

Figure 4.43: Voltage profile and Power losses of the IEEE 69 bus test system for penetration of Wind DG

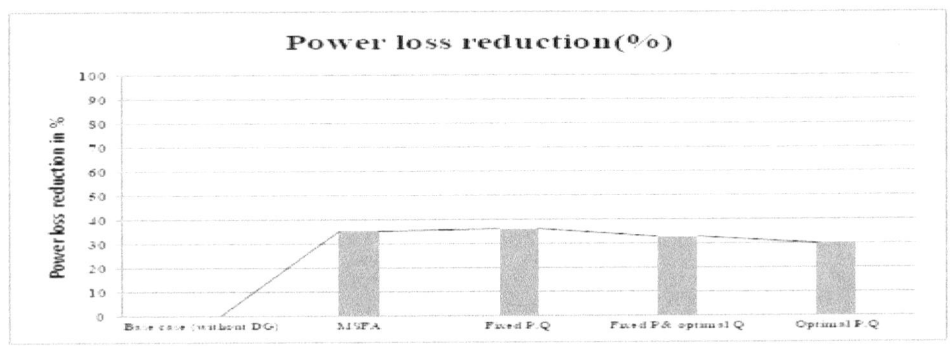

Figure 4.44: Power loss reduction in % of the IEEE 69 bus test system for scenario 1

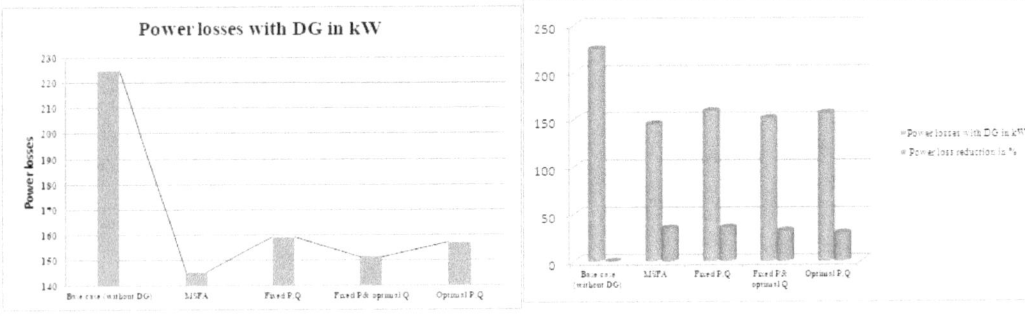

Figure 4.45: Power loss reduction in kW & % after introducing DG for MSFA and scenario 1 of MSFA-PSO

Case-2, Scenario-2: 20 % integration of a wind DG

The results from developed MSFA-PSO algorithm for allocation of a wind DG unit for scenario-1 are tabulated in table below. The analysis is performed for three different cases which is further compared to the previously calculated value of the modified shuffled frog leap algorithm. The first technique allocates the DG according to the fixed value of active and reactive power of the introduced wind generator in the system. Here, the system losses are reduced to 80.75 kW from 224.98 kW and percentage losses are reduced to 64.1. The next technique allocates the DG according to the fixed value of active power and optimal value of the reactive power which is calculated by the algorithm. Here, the system losses reduced are to 86.81 kW from 224.98 kW and percentage losses are reduced to 61.41. The last approach allocates the DG according to the optimal value of the active and reactive power which is calculated by the algorithm. Here, the system losses reduced are to 88.93 kW from 224.98 kW and percentage losses are reduced to 60.47. Minimum voltage before DG placement is 0.90918 p.u at bus 65 and after DG placement the minimum voltage obtained from the best of the three approaches is observed to be 0.95801 p.u.at bus 65. Figures show the graphs of voltage and power losses at different buses before and after integration of 20% of wind DG.

Table 4.35: Allocation of wind DG for scenario 2

Particulars	MSFA 20% wind DG	MSFA-PSO 20% wind DG (fixed P,Q)	MSFA-PSO 20% wind DG (Fixed P, Optimal Q within 20%)	MSFA-PSO wind DG (Optimal P, Q within 20%)
1: OLDG	Bus 63	Bus 64	Bus 61	Bus 61
2: ODGS	760 kW, -538 kVAR	760 kW, -538 kVAR	760 kW, -369.12 kVAR	749.18 kW, -453.21 kVAR
3: BCPL (kW)	224.98	224.98	224.98	224.98
4: PLDG (kW)	89.23	80.75	86.812	88.93
5: PLR (%)	60.33	64.1	61.41	60.47
6: Min V w/o DG (p.u.)	Bus 65	Bus 61	Bus 65	Bus 65
	0.90918	0.90918	0.90918	0.90918
7: Min V W DG (p.u.)	Bus 27	Bus 65	Bus 65	Bus 65

	0.95212	0.94766	0.95801	0.94766

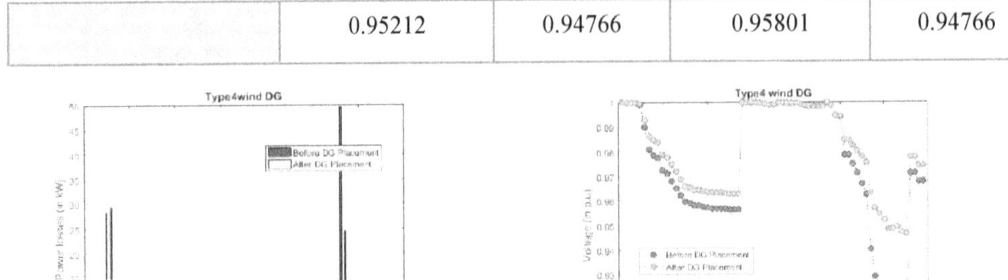

Figure 4.46: Voltage profile and Power losses of the IEEE 69 bus test system for penetration of

Wind DG

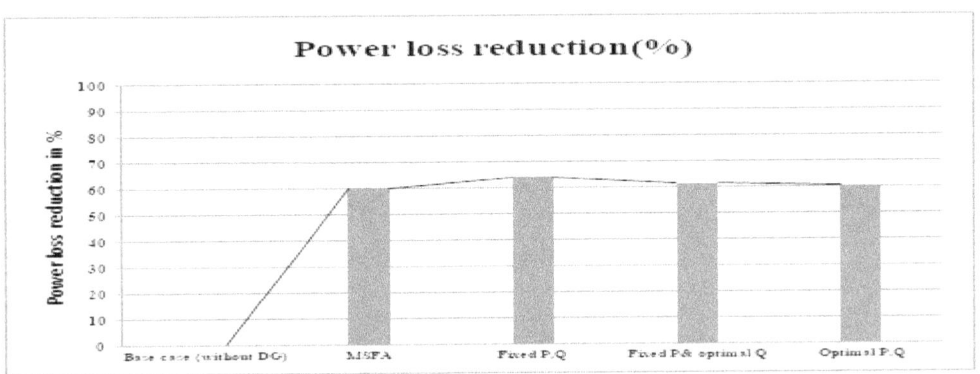

Figure 4.47: Power loss reduction in % of the IEEE 69 bus test system for scenario 2

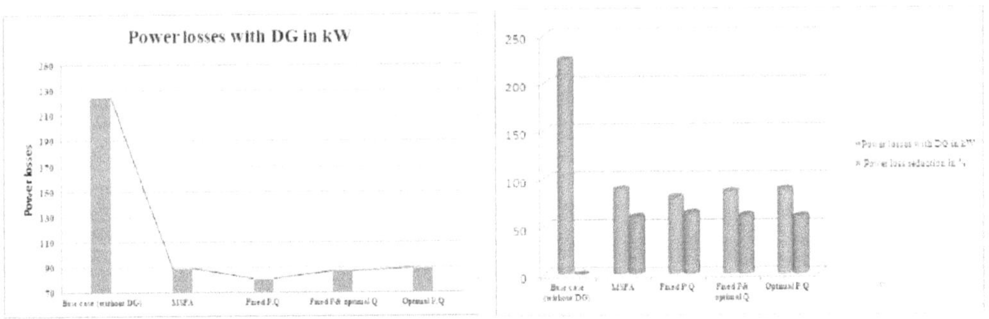

Figure 4.48: Power loss reduction in kW and % of the IEEE 69 bus test system for scenario 2

Case-2, Scenario-3: 30 % integration of a wind DG

The results from developed MSFA-PSO algorithm for allocation of a wind DG unit for scenario-1 are tabulated in table below. The analysis is performed for three different cases which is further compared to the previously calculated value of the modified shuffled frog leap algorithm. The first technique allocates the DG according to the fixed value of active and reactive power of the introduced wind generator in the system. Here, the system losses are reduced to 50.93 kW from 224.98 kW and percentage losses are reduced to 77.36. The next technique allocates the DG according to the fixed value of active power and optimal value of the reactive power which is calculated by the algorithm. Here, the system losses reduced are to 53.31 kW from 224.98 kW and percentage losses are reduced to 76.30. The last approach allocates the DG according to the optimal value of the active and reactive power which is calculated by the algorithm. Here, the system losses reduced are to 48.69 kW from 224.98 kW and percentage losses are reduced to 78.35. Minimum voltage before DG placement is 0.90918 p.u at bus 65 and after DG placement the minimum voltage obtained from the best of the three approaches is observed to be 0.9657 p.u.at bus 65. Figures shows the graphs of voltage and power losses at different buses before and after integration of 30% of wind DG.

Table 4.36: Allocation of wind DG for scenario 3

Particulars	MSFA 30% wind DG	MSFA-PSO 30% wind DG (fixed P,Q)	MSFA-PSO 30% wind DG (Fixed P, Optimal Q within 30%)	MSFA-PSO wind DG (Optimal P, Q within 30%)
1: OLDG	Bus 69	Bus 61	Bus 61	Bus 61
2: ODGS	1140 kW, -807 kVAR	1140 KW, -807 kVAR	1140 kW, - 646.99 kVAR	1019 kW, -563.9 kVAR
3: BCPL (kW)	224.98	224.98	224.98	224.98
4: PLDG (kW)	51.31	50.93	53.31	48.69
5: PLR (%)	77.19	77.36	76.30	78.35
6: Min V w/o DG (p.u.)	Bus 65	Bus 65	Bus 65	Bus 65
	0.90918	0.90918	0.90918	0.90918
7: Min V W DG (p.u.)	Bus 65	Bus 65	Bus 65	Bus 65

	0.95891	0.9657	0.95976	0.9657

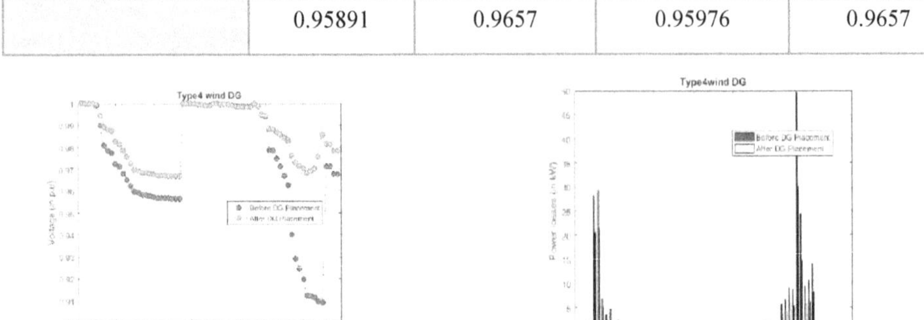

Figure 4.49: Voltage profile and Power losses of the IEEE 69 bus test system for penetration of Wind DG – scenario 3

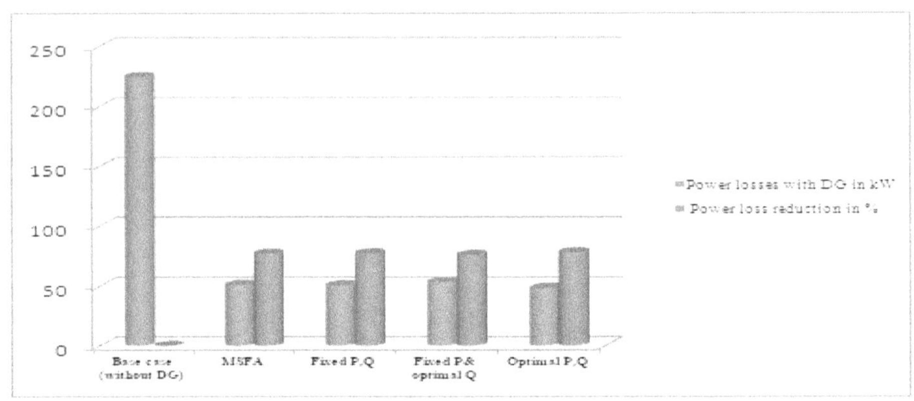

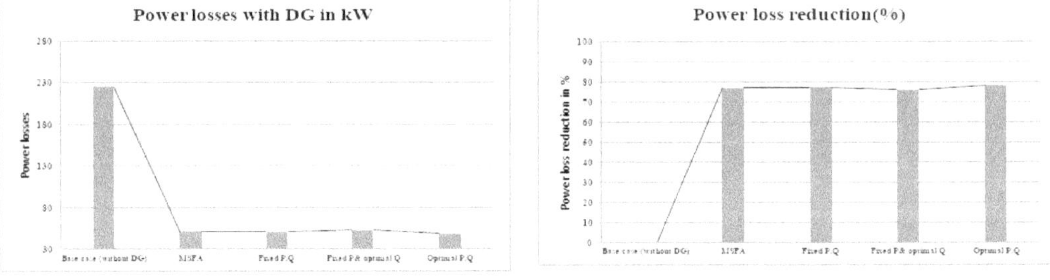

Figure 4.50: Power loss reduction in kW and % of the IEEE 69 bus test system for scenario 3

Case-2, Scenario-4: 40 % integration of a wind DG

The results from developed MSFA-PSO algorithm for allocation of a wind DG unit for scenario-1 are tabulated in table below. The analysis is performed for three different cases which is further compared to the previously calculated value of the modified shuffled frog leap

160

algorithm. The first technique allocates the DG according to the fixed value of active and reactive power of the introduced wind generator in the system. Here, the system losses are reduced to 33.12 kW from 224.98 kW and percentage losses are reduced to 85.27. The next technique allocates the DG according to the fixed value of active power and optimal value of the reactive power which is calculated by the algorithm. Here, the system losses reduced are to 34.12 kW from 224.98 kW and percentage losses are reduced to 84.83. The last approach allocates the DG according to the optimal value of the active and reactive power which is calculated by the algorithm. Here, the system losses reduced are to 29.78 kW from 224.988 kW and percentage losses are reduced to 86.76. Minimum voltage before DG placement is 0.90918 p.u at bus 65 and after DG placement the minimum voltage obtained from the best of the three approaches is observed to be 0.9789 p.u.at bus 27. Figures shows the graphs of voltage and power losses at different buses before and after integration of 40% of wind DG.

Table 4.37: Allocation of wind DG for scenario 4

Particulars	MSFA 40% wind DG	MSFA-PSO 40% wind DG (fixed P, Q)	MSFA-PSO 40% wind DG (Fixed P, Optimal Q within 40%)	MSFA-PSO wind DG (Optimal P, Q within 40%)
1: OLDG	Bus 61	Bus 62	Bus 62	Bus 62
2: ODGS	1520kW -1053.7 kVAR	1520 kW, -1053.7 kVAR	1520 kW, -989.89 kVAR	1329.29 kW, -992.69 kVAR
3: BCPL (kW)	224.98	224.98	224.98	224.98
4: PLDG (kW)	32.32	33.12	34.12	29.78
5: PLR (%)	85.63	85.27	84.83	86.76
6: Min V w/o DG (p.u.)	Bus 65	Bus 65	Bus 65	Bus 65
	0.90918	0.90918	0.90918	0.90918
7: Min V W DG (p.u.)	Bus 27	Bus 27	Bus 27	Bus 27
	0.9699	0.96231	0.9523	0.9789

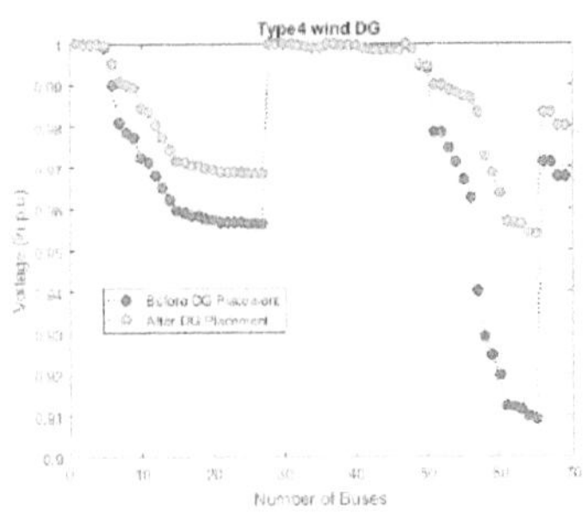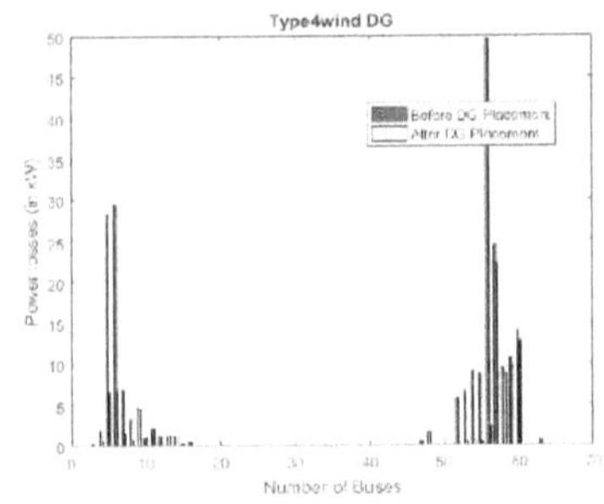

Figure 4.51: Voltage profile and Power losses of the IEEE 69 bus test system for penetration of Wind DG – scenario 4

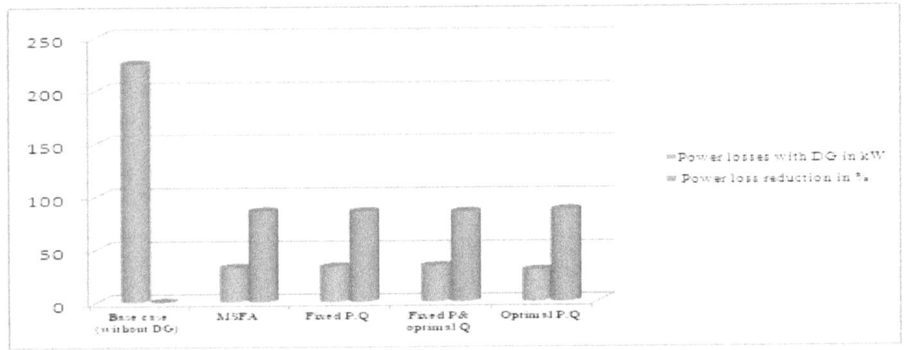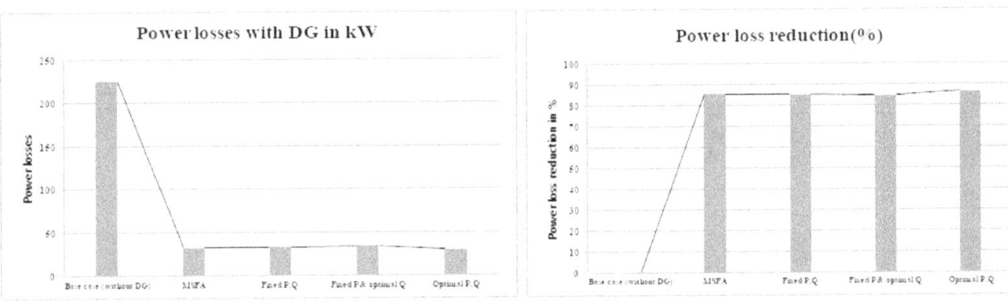

Figure 4.52: Power loss reduction in kW and % of the IEEE 69 bus test system for scenario 4

Case-2, Scenario-5: 50 % integration of a wind DG

The results from developed MSFA-PSO algorithm for allocation of a wind DG unit for scenario-1 are tabulated in table below. The analysis is performed for three different cases which is further compared to the previously calculated value of the modified shuffled frog leap algorithm. The first technique allocates the DG according to the fixed value of active and reactive power of the introduced wind generator in the system. Here, the system losses are reduced to 23.32 kW from 224.98 kW and percentage losses are reduced to 89.63. The next technique allocates the DG according to the fixed value of active power and optimal value of the reactive power which is calculated by the algorithm. Here, the system losses reduced are to 24.98 kW from 224.98 kW and percentage losses are reduced to 88.89.

Here, the system losses reduced are to 22.59 kW from 224.98 kW and percentage losses are reduced to 89.95. Minimum voltage before DG placement is 0.90918 p.u at bus 65 and after DG placement the minimum voltage obtained from the best of the three approaches is observed to be 0.9893 p.u.at bus 27.

Table 4.38: Allocation of wind DG for scenario 5

Particulars	MSFA 50% wind DG	MSFA-PSO 50% wind DG (fixed P, Q)	MSFA-PSO 50% wind DG (Fixed P, Optimal Q within 50%)	MSFA-PSO wind DG (Optimal P, Q within 50%)
1: OLDG	Bus 63	Bus 61	Bus 61	Bus 61
2: ODGS	1900 kW, -1300.2 kVAR	1900 kW, -1300.2 kVAR	1900 kW, -1196.2 kVAR	1723.1 kW, -1183.3 kVAR
3: BCPL (kW)	224.98	224.98	224.98	224.98
4: PLDG (kW)	25.31	23.32	24.98	22.59
5: PLR (%)	88.75	89.63	88.89	89.95
6: Min V w/o DG (p.u.)	Bus 65	Bus 65	Bus 65	Bus 65
	0.90918	0.90918	0.90918	0.90918
7: Min V W DG (p.u.)	Bus 27	Bus 27	Bus 27	Bus 27
	0.96982	0.97293	0.97321	0.9893

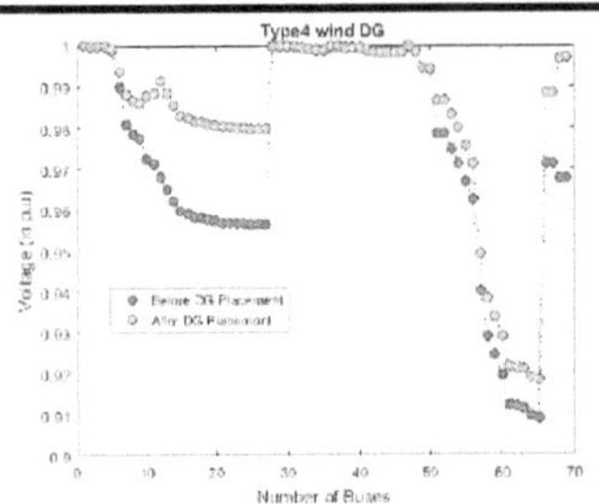
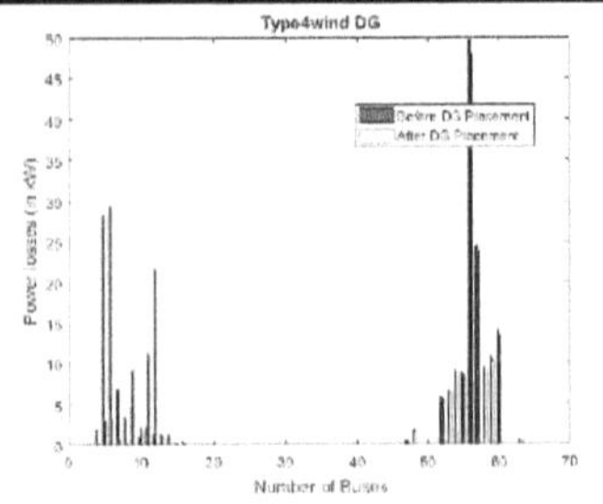

Figure 4.53: Voltage profile and Power losses of the IEEE 69 bus test system for penetration of Wind DG – scenario 5

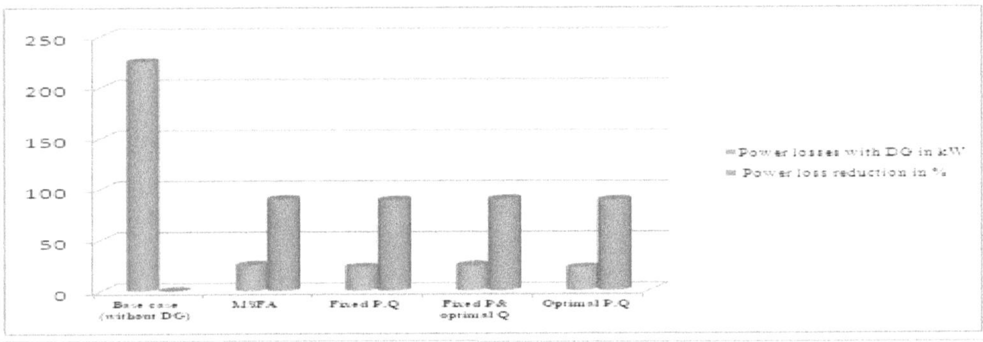

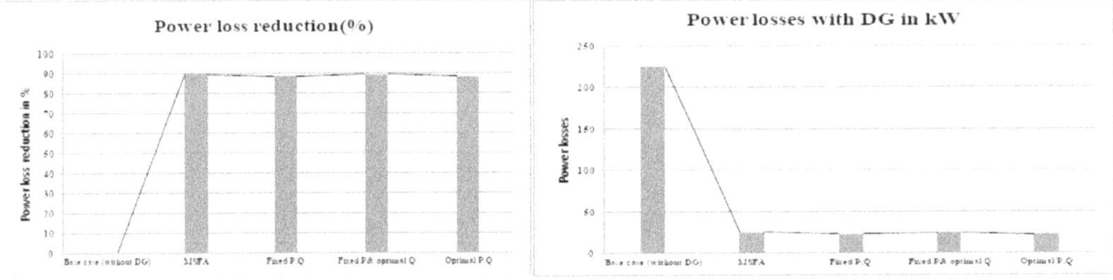

Figure 4.54: Power loss reduction in kW and % of the IEEE 69 bus test system for scenario 5

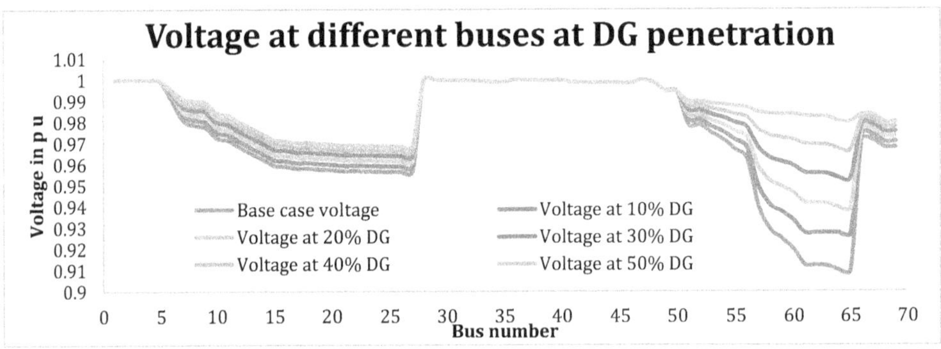

Figure 4.55: Voltage profile of the IEEE 69 bus test system for DG penetration from 10% to 50% of Wind Penetration

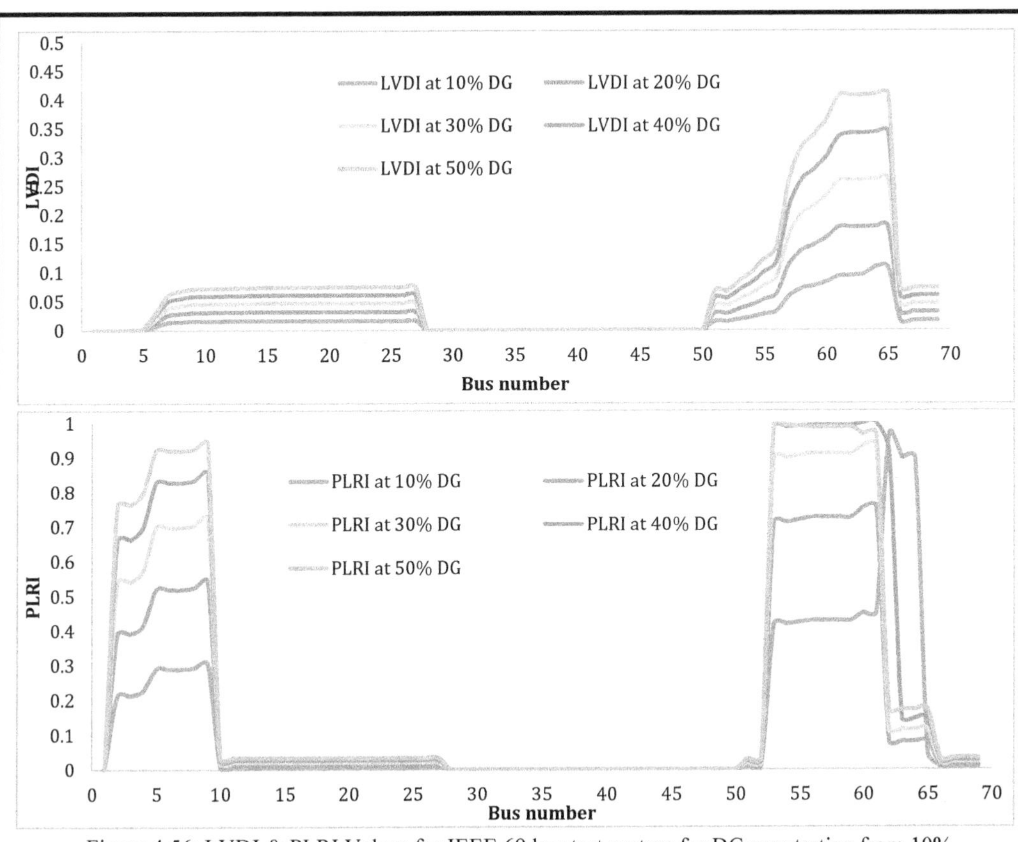

Figure 4.56: LVDI & PLRI Values for IEEE 69 bus test system for DG penetration from 10% to 50% of Wind Penetration

Table 4.39: WSOF values at different buses for integration of solar DG at different penetration level

Bus number	WSOF at 10%	WSOF at 20%	WSOF at 30%	WSOF at 40%	WSOF at 50%
1	0	0	0	0	0
2	0.085004	0.156599	0.21681	0.265324	0.30534
3	0.085031	0.156651	0.216836	0.265376	0.305418
4	0.090963	0.166785	0.229667	0.279544	0.319765
5	0.11612	0.208456	0.280374	0.332594	0.369447
6	0.119803	0.215661	0.290964	0.346328	0.386267
7	0.123843	0.223462	0.30235	0.36104	0.404161
8	0.126332	0.227796	0.307927	0.367374	0.410785
9	0.131555	0.236301	0.318019	0.377527	0.419715
10	0.012523	0.024419	0.035795	0.046404	0.056683

11	0.012572	0.024491	0.035883	0.046498	0.056785
12	0.012667	0.024635	0.036097	0.046797	0.057161
13	0.012748	0.024791	0.036329	0.047087	0.057492
14	0.012802	0.024923	0.036503	0.047325	0.057786
15	0.012801	0.025001	0.036658	0.047491	0.058016
16	0.012865	0.025055	0.036711	0.047586	0.058095
17	0.012902	0.025079	0.036763	0.047653	0.058188
18	0.012907	0.025107	0.036758	0.047659	0.058184
19	0.012855	0.025099	0.036767	0.047687	0.058225
20	0.012913	0.025128	0.036817	0.04772	0.058266
21	0.01291	0.025148	0.036852	0.047739	0.05831
22	0.012917	0.025138	0.036833	0.047774	0.058316
23	0.012922	0.025142	0.036838	0.047744	0.058315
24	0.012936	0.025196	0.036866	0.047798	0.05832
25	0.012893	0.025145	0.036846	0.047761	0.058335
26	0.012929	0.025157	0.036857	0.047769	0.058344
27	0.01293	0.025152	0.036853	0.047803	0.058344
28	0	3.76E-05	6.36E-05	7.52E-05	0.000101
29	2.61E-05	6.76E-05	9.37E-05	9.37E-05	0.000135
30	3.29E-05	6.59E-05	7.27E-05	9.88E-05	0.000132
31	2.61E-05	5.21E-05	5.21E-05	0.000117	0.000143
32	3.38E-05	6.77E-05	9.38E-05	0.000102	0.000135
33	5.86E-05	5.86E-05	8.47E-05	0.000111	0.000111
34	2.61E-05	5.22E-05	5.22E-05	7.82E-05	0.000104
35	0	3.44E-05	6.89E-05	6.89E-05	0.000103
36	2.61E-05	2.61E-05	8.06E-05	0.000107	0.000107
37	0	2.61E-05	5.21E-05	7.82E-05	0.000105
38	2.61E-05	2.61E-05	7.52E-05	0.000101	0.000101
39	3.41E-05	6.81E-05	7.61E-05	0.000102	0.000136
40	2.61E-05	7.23E-05	7.23E-05	9.84E-05	9.84E-05
41	2.61E-05	6.04E-05	6.86E-05	9.47E-05	0.000129
42	2.61E-05	5.22E-05	7.21E-05	9.82E-05	9.82E-05
43	2.61E-05	6.72E-05	6.72E-05	9.33E-05	0.000134

44	2.61E-05	6E-05	9.39E-05	0.000102	0.000128
45	6.58E-06	3.93E-05	7.19E-05	7.85E-05	0.000105
46	2.61E-05	5.22E-05	7.83E-05	0.00011	0.000136
47	6.93E-05	0.000139	0.000182	0.000234	0.000277
48	6.59E-05	0.000132	0.000198	0.000238	0.000303
49	7.69E-05	0.000103	0.00018	0.000232	0.000283
50	5.24E-05	0.000139	0.000192	0.000252	0.000305
51	0.011713	0.02279	0.033409	0.043317	0.052929
52	0.011698	0.022788	0.033414	0.043307	0.052912
53	0.17857	0.306807	0.391517	0.436334	0.448123
54	0.18105	0.311409	0.397923	0.444288	0.457385
55	0.18609	0.320007	0.408781	0.456243	0.469442
56	0.190986	0.328371	0.419371	0.467949	0.48134
57	0.206586	0.358507	0.463235	0.524516	0.550029
58	0.214491	0.373721	0.485278	0.552872	0.584364
59	0.217591	0.379686	0.493908	0.563953	0.597752
60	0.23005	0.398536	0.513837	0.580361	0.606563
61	0.235833	0.409606	0.529782	0.600777	0.631256
62	0.442542	0.139184	0.201642	0.571018	0.313349
63	0.417718	0.139287	0.201773	0.263405	0.313703
64	0.424858	0.139695	0.202394	0.264175	0.31451
65	0.084671	0.139847	0.20261	0.264464	0.314849
66	0.012528	0.024386	0.035745	0.046347	0.05661
67	0.01251	0.024375	0.035733	0.04635	0.0566
68	0.012641	0.024576	0.036023	0.046684	0.057005
69	0.012631	0.024573	0.035995	0.046674	0.057003

Case-3, Scenario-1: 10 % integration of SOLAR, WIND Hybrid DG

The results from developed MSFA-PSO algorithm for allocation of Hybrid DG unit for scenario-1 are tabulated in table 4.40. The system losses reduced to 141.41 kW from 224.98 kW and percentage losses are reduced by 37.14%. Minimum voltage before DG placement is 0.90918 p.u at bus 65 and after DG placement it the minimum voltage obtained to be 0.93669 p.u.at bus 62.Figures 4.57 and 4.58 shows the variation of voltage and power losses at different buses before and after integration of 10% of hybrid DG.

Table 4.40: Allocation of Hybrid DG for scenario 1

Particulars	MSFA Hybrid DG	MSFA-PSO Hybrid DG
1: OLDG	Bus 65 Bus 62	Bus 59 Bus 62
2: ODGS	280 kW, 380 kW, -23.39 kVAR	280 kW, 380 kW, -23.39 kVAR
3: BCPL (kW)	224.98	224.98
4: PLDG (kW)	151.32	141.41
5: PLR (%)	32.74	37.14
6: Min V w/o DG (p.u.)	Bus 65	Bus 65
	0.90918	0.90918
7: Min V W DG (p.u.)	Bus 65	Bus 62
	0.93225	0.93669

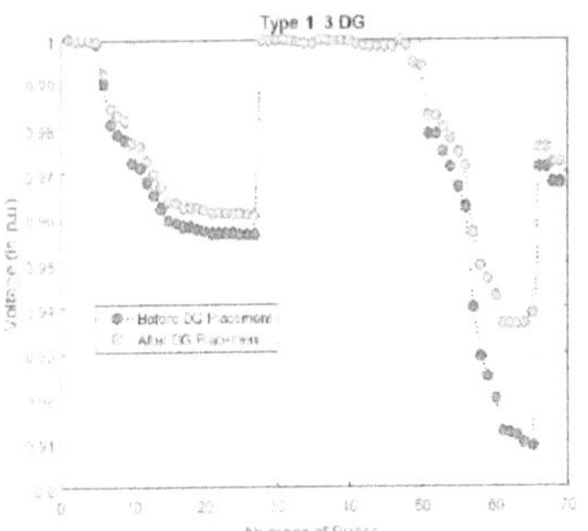
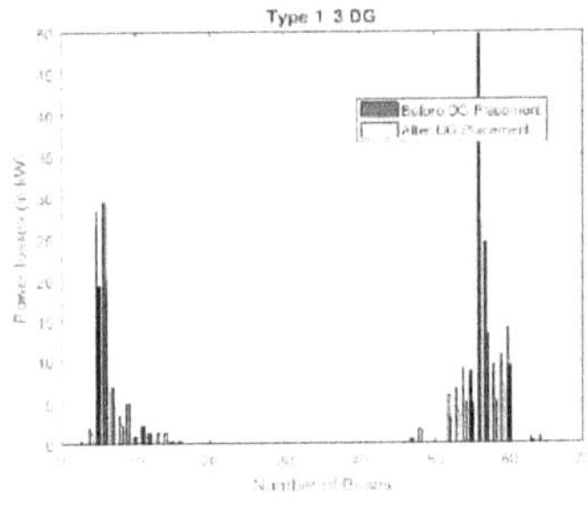

Figure 4.57: Voltage profile and Power losses of the IEEE 69 bus test system for penetration of Hybrid DG – scenario 1

Case-3, Scenario-2: 20 % integration of SOLAR, WIND Hybrid DG

The results from developed MSFA-PSO algorithm for allocation of Hybrid DG unit for scenario-2 are tabulated in table 4.41 The system losses reduced to 59.04 kW from 224.98 kW and percentage losses are reduced by 73.75%. Minimum voltage before DG placement is

0.90918 p.u at bus 65 and after DG placement it the minimum voltage obtained to be 0.9675 p.u.at bus 27. Figure 4.58 shows the variation of voltage and power losses at different buses before and after integration of 20% of hybrid DG.

Table 4.41: Allocation of Hybrid DG for scenario 2

Particulars	MSFA Hybrid DG	MSFA-PSO Hybrid DG
1: OLDG	Bus 65, 63	Bus 61, 62
2: ODGS	760 kW, 760 kW, -345.7 kVAR	760 kW, 760 kW, -345.7 kVAR
3: BCPL (kW)	224.98	224.98
4: PLDG (kW)	62.13	59.04
5: PLR (%)	72.38	73.75
6: Min V w/o DG (p.u.)	Bus 65	Bus 65
	0.90918	0.90918
7: Min V W DG (p.u.)	Bus 65	Bus 27
	0.94561	0.9675

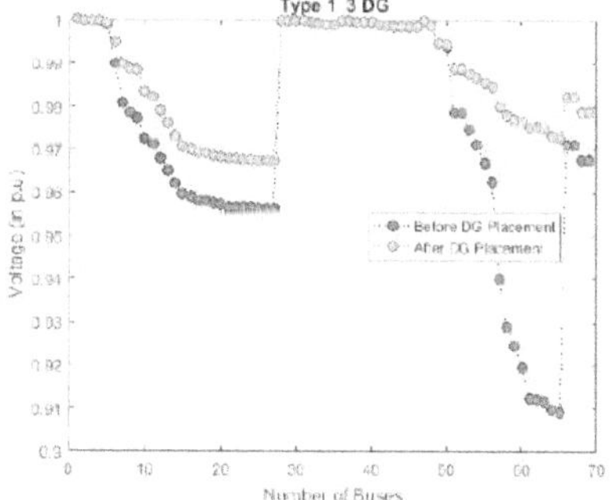
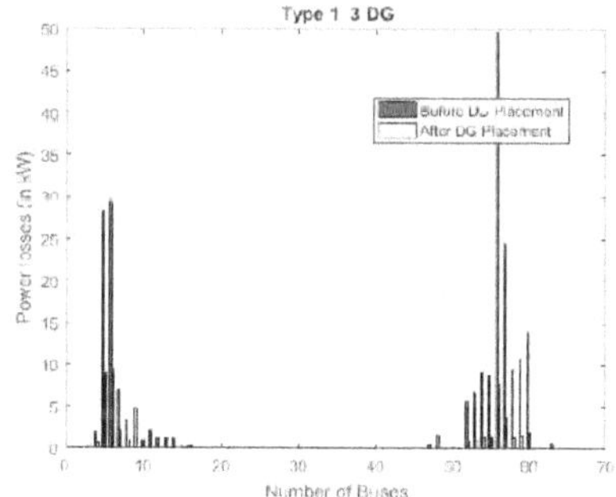

Figure 4.58: Voltage profile and Power losses of the IEEE 69 bus test system for penetration of Hybrid DG – scenario 2

Case-3, Scenario-3: 30 % integration of SOLAR, WIND Hybrid DG

The results from developed MSFA-PSO algorithm for allocation of Hybrid DG unit for scenario-3 are tabulated in table 4.42. The system losses reduced to 71.71 kW from 224.98 kW and percentage losses are reduced by 68.36%. Minimum voltage before DG placement is 0.90918 p.u at bus 65 and after DG placement it the minimum voltage obtained to be 0.96416 p.u.at bus 65. Figure 4.59 shows the variation of voltage and power losses at different buses before and after integration of 20% of hybrid DG.

Table 4.42: Allocation of Hybrid DG for scenario 3

Particulars	MSFA Hybrid DG	MSFA-PSO Hybrid DG
1: OLDG	Bus 63, 65	Bus 63, 10
2: ODGS	1140 kW, 1140 kW, - 785.2 kVAR	1140 kW, 1140 kW, -785.2 kVAR
3: BCPL (kW)	224.98	224.98
4: PLDG (kW)	74.23	71.17
5: PLR (%)	67	68.36
6: Min V w/o DG (p.u.)	Bus 65	Bus 65
	0.90918	0.90918
7: Min V W DG (p.u.)	Bus 27	Bus 27
	0.95921	0.96416

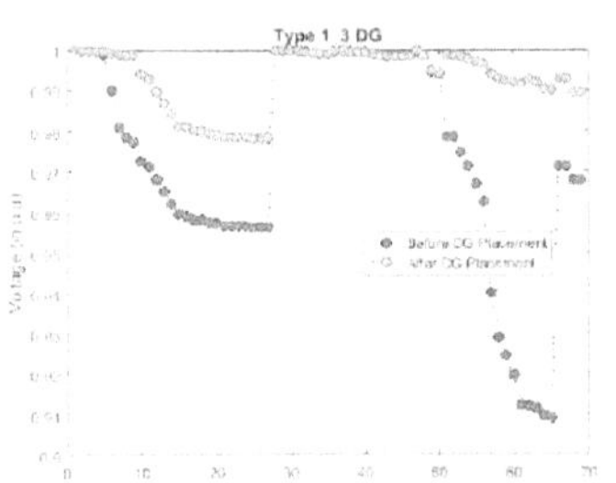
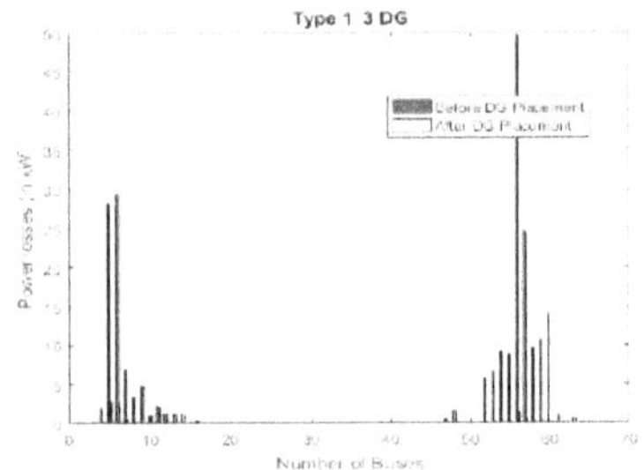

Figure 4.59: Voltage profile and Power losses of the IEEE 69 bus test system for penetration of Hybrid DG – scenario 3

Case-3, Scenario-4: 40 % integration of SOLAR, WIND Hybrid DG

The results from developed MSFA-PSO algorithm for allocation of Hybrid DG unit for scenario-4 are tabulated in table 4.43 The system losses reduced to 26.913 kW from 224.98 kW and percentage losses are reduced by 88.03%. Minimum voltage before DG placement is 0.90918 p.u at bus 65 and after DG placement it the minimum voltage obtained to be 0.9782 p.u.at bus 27. Figure 4.60 shows the variation of voltage and power losses at different buses before and after integration of 40% of hybrid DG.

Table 4.43: Allocation of Hybrid DG for scenario 4

Particulars	MSFA Hybrid DG	MSFA-PSO Hybrid DG
1: OLDG	Bus 62,65	Bus 63, 10
2: ODGS	1520 kW, 1520 kW, - 854.02 kVAR	1520 kW, 1520 kW, -854.02 kVAR
3: BCPL (kW)	224.98	224.98
4: PLDG (kW)	29.23	26.91
5: PLR (%)	87.01	88.03
6: Min V w/o DG (p.u.)	Bus 65	Bus 65
	0.90918	0.90918
7: Min V W DG (p.u.)	Bus 27	Bus 27
	0.9545	0.9782

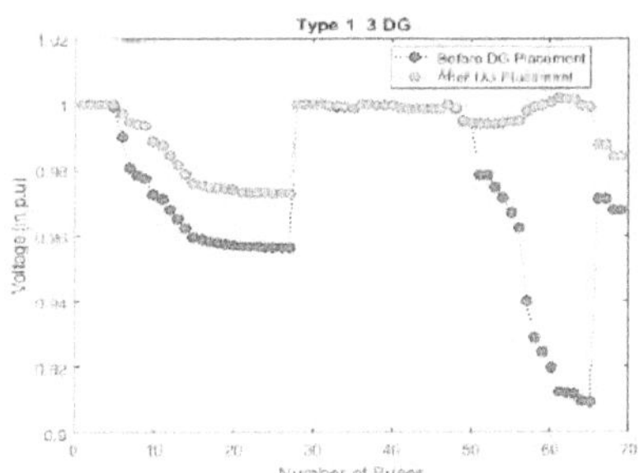
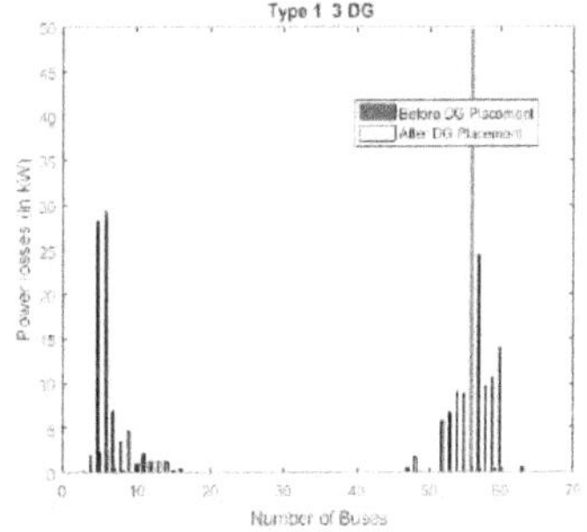

Figure 4.60: Voltage profile and Power losses of the IEEE 69 bus test system for penetration of Hybrid DG – scenario 4

Case-3, Scenario-5: 50 % integration of SOLAR, WIND Hybrid DG

The results from developed MSFA-PSO algorithm for allocation of Hybrid DG unit for scenario-5 are tabulated in table 4.44. The system losses reduced to 23.286 kW from 224.98 kW and percentage losses are reduced by 89.65%. Minimum voltage before DG placement is 0.90918 p.u at bus 65 and after DG placement it the minimum voltage obtained to be 0.9792 p.u.at bus 27. Figures 4.61 & 4.62 shows the variation of voltage and power losses at different buses before and after integration of 50% of hybrid DG.

Table 4.44: Allocation of Hybrid DG for scenario 5

Particulars	MSFA Hybrid DG	MSFA-PSO Hybrid DG
1: OLDG	Bus 63 Bus 49	Bus 47 Bus 61
2: ODGS	1900 kW, 1900 kW, - 1287 kVAR	1900 kW, 1900 kW, -1287 kVAR
3: BCPL (kW)	224.98	224.98
4: PLDG (kW)	25.65	23.28
5: PLR (%)	88.59	89.65
6: Min V w/o DG (p.u.)	Bus 65	Bus 65
	0.90918	0.90918
7: Min V W DG (p.u.)	Bus 27	Bus 27
	0.9569	0.9792

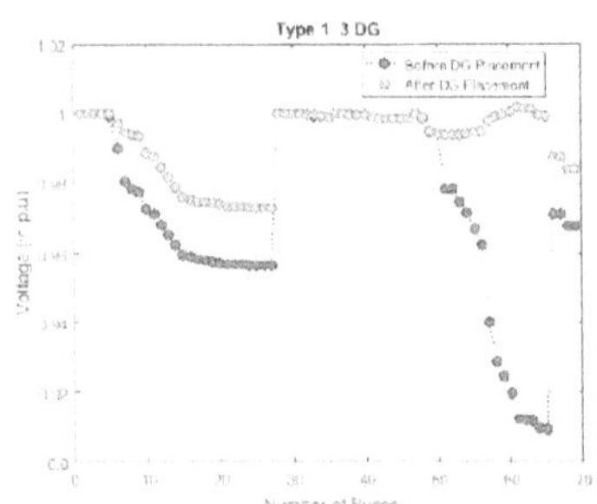
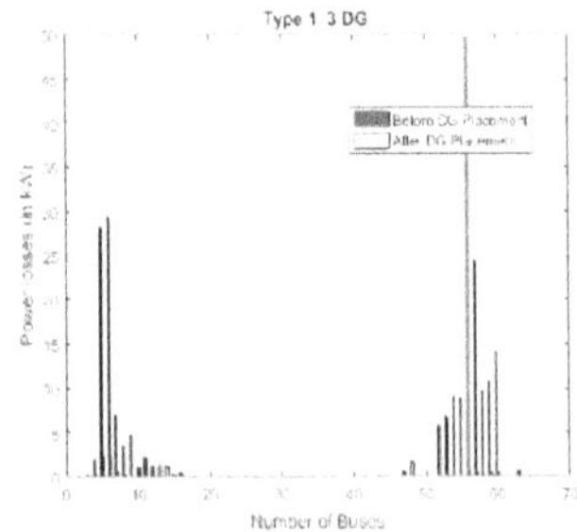

Figure 4.61: Voltage profile and Power losses of the IEEE 69 bus test system for penetration of Hybrid DG – scenario 5

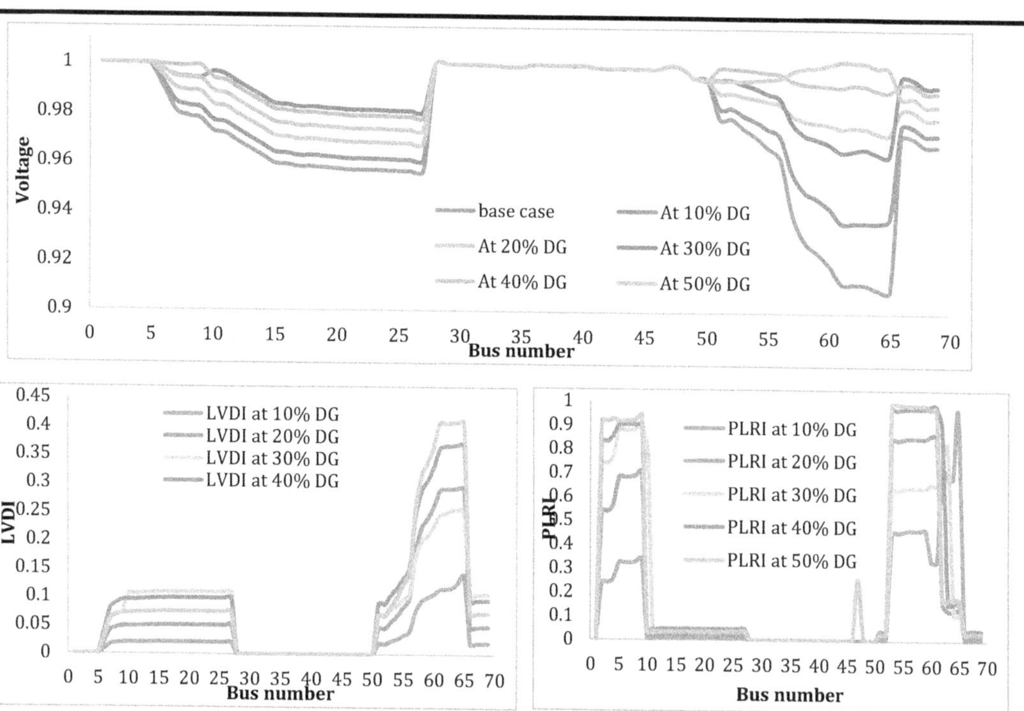

Figure 4.62: Voltage profile, LVDI & PLRI Values for IEEE 69 bus test system for DG penetration from 10% to 50% of Hybrid Penetration

Table 4.45: WSOF values at different buses for integration of hybrid DG at different penetration level

Bus number	WSOF at 10%	WSOF at 20%	WSOF at 30%	WSOF at 40%	WSOF at 50%
1	0	0	0	0	0
2	0.095915	0.214626	0.294992	0.332336	0.366419
3	0.095928	0.21466	0.295044	0.332414	0.366498
4	0.102324	0.226668	0.308248	0.342184	0.370155
5	0.129416	0.273322	0.352781	0.362582	0.369003
6	0.134051	0.284725	0.369967	0.384807	0.385772
7	0.139136	0.296948	0.388242	0.408265	0.40364
8	0.141977	0.302452	0.394645	0.413858	0.41025

9	0.147593	0.311608	0.402439	0.415411	0.419214
10	0.015628	0.038256	0.377677	0.07438	0.056649
11	0.015672	0.038343	0.082425	0.074533	0.056751
12	0.015757	0.038598	0.082933	0.074997	0.057116
13	0.015857	0.038812	0.083461	0.075453	0.057465
14	0.01594	0.039009	0.083833	0.075814	0.057732
15	0.015998	0.039156	0.084204	0.076145	0.05799
16	0.016032	0.039217	0.084312	0.076235	0.058051
17	0.016049	0.039286	0.084434	0.076331	0.058149
18	0.016055	0.039279	0.084426	0.076356	0.058157
19	0.016062	0.039281	0.084493	0.07639	0.05816
20	0.016066	0.039341	0.084556	0.076449	0.058227
21	0.016069	0.03935	0.084624	0.076523	0.058256
22	0.016104	0.039389	0.08462	0.076508	0.058282
23	0.016103	0.039387	0.084621	0.076515	0.058281
24	0.016128	0.039413	0.084677	0.07654	0.058293
25	0.016075	0.039374	0.084653	0.076536	0.058301
26	0.016081	0.039401	0.084652	0.076572	0.058301
27	0.016088	0.039415	0.084665	0.076571	0.058317
28	0	3.76E-05	7.52E-05	0.000101	0.000139
29	2.61E-05	6.76E-05	9.37E-05	0.000135	0.000161
30	3.29E-05	6.59E-05	9.88E-05	0.000132	0.000165
31	2.61E-05	5.21E-05	0.000117	0.000143	0.000169
32	3.38E-05	6.77E-05	0.000102	0.000135	0.000169

33	3.25E-05	8.47E-05	0.000111	0.000111	0.000169
34	2.61E-05	5.22E-05	7.82E-05	0.000104	0.000169
35	0	3.44E-05	6.89E-05	0.000103	0.000138
36	0	8.06E-05	0.000107	0.000107	0.000133
37	0	2.61E-05	7.82E-05	0.000105	0.000131
38	0	5.21E-05	0.000101	0.000101	0.000127
39	3.41E-05	6.81E-05	0.000102	0.000136	0.00017
40	2.61E-05	7.23E-05	9.84E-05	0.000124	0.000171
41	2.61E-05	6.04E-05	9.47E-05	0.000129	0.000163
42	2.61E-05	7.21E-05	9.82E-05	9.82E-05	0.00017
43	2.61E-05	6.72E-05	9.33E-05	0.000134	0.000161
44	2.61E-05	6E-05	0.000102	0.000128	0.000162
45	6.58E-06	3.93E-05	7.85E-05	0.000111	0.000144
46	2.61E-05	5.22E-05	0.00011	0.000136	0.000162
47	4.32E-05	0.000139	0.000234	0.000303	0.102638
48	5.91E-05	0.000165	0.000238	0.000303	0.000541
49	2.62E-05	0.000129	0.000232	0.000283	0.000515
50	5.24E-05	0.000166	0.000252	0.000305	0.000557
51	0.014575	0.035721	0.053815	0.069421	0.05288
52	0.014567	0.035701	0.053803	0.069429	0.052886
53	0.194868	0.369891	0.299221	0.44907	0.4482
54	0.197893	0.376659	0.304069	0.456756	0.45747
55	0.203663	0.387301	0.312214	0.468588	0.469544
56	0.209249	0.397683	0.320159	0.48016	0.48143

57	0.230554	0.447483	0.356098	0.534984	0.550016
58	0.241406	0.472575	0.374322	0.562517	0.584299
59	0.245658	0.482392	0.381469	0.573263	0.597661
60	0.197211	0.499548	0.396851	0.59092	0.606723
61	0.201963	0.516832	0.409122	0.610499	0.631157
62	0.357284	0.238765	0.454355	0.580671	0.312986
63	0.343908	0.225564	0.479867	0.282899	0.313094
64	0.349373	0.226252	0.1971	0.283739	0.314016
65	0.462039	0.226515	0.197316	0.284063	0.314351
66	0.015596	0.038219	0.082157	0.074292	0.056569
67	0.015592	0.038192	0.08216	0.074251	0.056548
68	0.015747	0.038497	0.082738	0.074826	0.056961
69	0.01574	0.038485	0.082737	0.074817	0.056966

Case-4, Scenario 1- : Optimal placement of a SOLAR DG

The results obtained from the developed MSFA-PSO algorithm for case-1, scenario-1 and comparison with SFA is tabulated in table.4.46. Total losses of the system are reduced to 83.20 kW from 224.98KW with percentage losses are reduced to 63.01%.

Table 4.46: Optimal allocation of Solar DG by MSFA-PSO and compared with MSFA

Particulars	MSFA Solar DG	MSFA-PSO Solar DG
1: OLDG	Bus 62	Bus 61
2: ODGS	1857.5 kW	1857.5 kW
3: BCPL (kW)	224.98	224.98
4: PLDG (kW)	84.80	83.20
5: PLR (%)	62.30	63.01

6: Min V w/o DG (p.u.)	Bus 65	Bus 65
	0.90918	0.90918
7: Min V W DG (p.u.)	Bus 27	Bus 27
	0.96424	0.96824

Case-4, Scenario 2- : optimal placement of two SOLAR DGs

The results obtained from the developed MSFA-PSO algorithm for case-4, scenario-2 and comparison with SFA is tabulated in table.4.47. Total losses of the system are reduced to 73.41 kW from 224.98 kW with percentage losses are reduced to 67.36%.

Table 4.47: Optimal allocation of two Solar DGs by MSFA-PSO and compared with MSFA

Particulars	MSFA Solar, 2DGs	MSFA-PSO Solar, 2 DGs
1: OLDG	Bus 61 Bus 18	Bus 61 Bus 22
2: ODGS	1857.5 kW, 19883 kW	1857.5 kW, 19883 kW
3: BCPL (kW)	224.98	224.98
4: PLDG (kW)	72.25	73.41
5: PLR (%)	67.88	67.36
6: Min V w/o DG (p.u.)	Bus 65	Bus 65
	0.90918	0.90918
7: Min V W DG (p.u.)	Bus 65	Bus 65
	0.97448	0.9853

Case-5, Scenario 1- : optimal placement of WIND DGs

The results obtained from the developed MSFA-PSO algorithm for case-5, scenario-1 and comparison with SFA is tabulated in table.4.48. Total losses of the system are reduced to 84.48 kW from 224.98 kW with percentage losses are reduced to 62.45%.

Table 4.48: Optimal allocation of wind DGs by MSFA-PSO and compared with SFA

Particulars	MSFA Wind DG	MSFA-PSO Wind DG
1: OLDG	Bus 64	Bus 61
2: ODGS	1589.2 kW, 1100 kVAR	1589.2 kW, - 1100 kVAR
3: BCPL (kW)	224.98	224.98
4: PLDG (kW)	85.32	84.48
5: PLR (%)	62.07	62.45
6: Min V w/o DG (p.u.)	Bus 65	Bus 65
	0.90918	0.90918
7: Min V W DG (p.u.)	Bus 27	Bus 27
	0.9676	0.9732

Case-5, Scenario 2- : optimal placement of two WIND DGs

The results obtained from the developed MSFA-PSO algorithm for case-5, scenario-2 and comparison with SFA is tabulated in table.4.51. Total losses of the system are reduced to 73.41 kW from 224.98 kW with percentage losses are reduced to 67.36%.

Table 4.49: Optimal allocation of two wind DGs by MSFA-PSO and compared with SFA

Particulars	MSFA 2 Wind DG	MSFA-PSO 2 Wind DG
1: OLDG	Bus 37 Bus 62	Bus 10 Bus 63
2: ODGS	500 kW, 1460 kVAR 1360 kW 475 kVAR	500 kW, - 1460 kVAR 1360 kW, -475 kVAR
3: BCPL (kW)	224.98	224.98
4: PLDG (kW)	75.12	73.41
5: PLR (%)	66.61	67.36
6: Min V w/o DG (p.u.)	Bus 65	Bus 65

	0.90918	0.90918
7: Min V W DG (p.u.)	Bus 27	Bus 65
	0.9812	0.9853

Case-6, Scenario 1- : optimal placement of SOLAR, WIND Hybrid DGs

The results obtained from the developed MSFA-PSO algorithm for case-6, scenario-1 and comparison with SFA is tabulated in table.4.52. Total losses of the system are reduced to 35.36 kW from 224.98 kW with percentage losses are reduced to 84.28%.

Table 4.50: Optimal allocation of hybrid DGs by MSFA-PSO and compared with SFA

Particulars	MSFA Hybrid DG	MSFA-PSO Hybrid DG
1: OLDG	Bus 37 Bus 62	Bus 10 Bus 63
2: ODGS	1460 kW 1360 kW, - 475 kVAR	1460 kW 1360 kW, -475 kVAR
3: BCPL (kW)	224.98	224.98
4: PLDG (kW)	37.23	35.36
5: PLR (%)	83.45	84.28
6: Min V w/o DG (p.u.)	Bus 65	Bus 65
	0.90918	0.90918
7: Min V W DG (p.u.)	Bus 27	Bus 27
	0.96661	0.96726

4.4.2 Case Study 2: 32 bus practical radial distribution network

The proposed method is tested on the 32 bus radial distribution network (Bhuvaneshwari feeder, Mysuru, India) a practical system. The loads are assumed to be constant with the voltage level of 11kv and 20 MVA Base. 3715kW and 2300kVar is the total active and reactive power of the system respectively. Fig 3.7 shows the diagram of the network considered. The basic data of the network that is the line data and the branch data are given in appendix A. The nodes with the highest amount of deviation are identified and tabulated in the table. The base case load flow is obtained where the power losses without the penetration of the renewable DG's are obtained to be 1.1109kW.

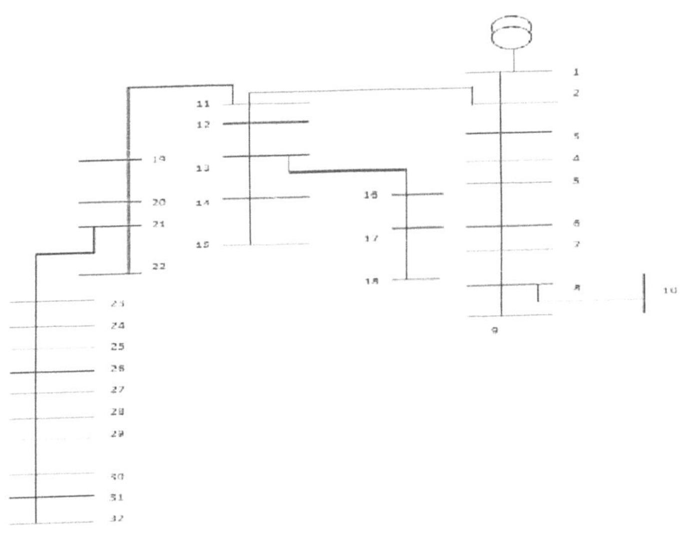

Fig 4.63: Single line diagram of 32 bus practical test system.

Case-1, Scenario-1: 10 % integration of a SOLAR DG
The results from developed MSFA-PSO algorithm for allocation of a Solar DG unit for scenario-1 are tabulated in table 4.53. The system losses reduced to 345.07 kW from 477.99 kW and percentage losses are reduced to 27.8 %. Minimum voltage before DG placement is 0.85952 p.u at bus 32 and after DG placement the minimum voltage obtained to be 0.90709 p.u.at bus 28.

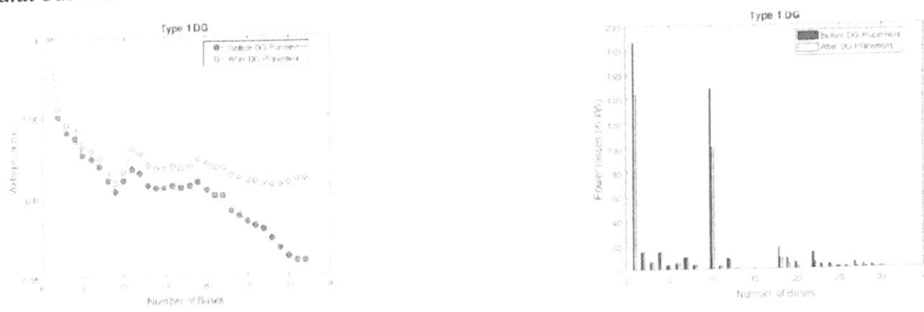

Fig 4.64: . Voltage profile and Power losses of the practical test system for penetration of solar DG – scenario 1

Table 4.53: Allocation of Solar DG for scenario 1

Particulars	MSFA 10%Solar DG	MSFA-PSO 10% Solar DG
1: OLDG	Bus 27	Bus 32
2: ODGS	174.3KW	174.3 KW,
3: BCPL (kW)	477.09	477.09
4: PLDG (kW)	349.29	345.07
5: PLR (%)	26.78	27.80
6: Min V w/o DG (p.u.)	Bus 32	Bus 32
	0.85952	0.85952
7: Min V W DG (p.u.)	Bus 09	Bus 28
	0.90012	0.90709

Case-1, Scenario-2: 20 % integration of a SOLAR DG:

The results from developed MSFLA-PSO algorithm for allocation of a Solar DG unit for scenario-2 are tabulated in table 4.54. The system losses reduced to 343.63 kW from 477.99 kW and percentage losses are reduced to 28.1 %. Minimum voltage before DG placement is 0.85952 p.u at bus 32 and after DG placement the minimum voltage obtained to be 0.90586 p.u.at bus 32.Figures shows the variation of voltage and power losses at different buses before and after integration of 20% of Solar DG.

Table 4.54: Allocation of Solar DG for scenario 2

Particulars	MSFA 20% SOLAR DG	MSFA-PSO 20% SOLAR DG
1: OLDG	Bus 32	Bus 30
2: ODGS	743kW	743 kW
3: BCPL (kW)	477.09	477.09
4: PLDG (kW)	351.31	343.63
5: PLR (%)	26.36	28.1
6: Min V w/o DG (p.u.)	Bus 32	Bus 32
	0.85952	0.85952
7: Min V W DG (p.u.)	Bus 32	Bus 32
	0.90236	0.90586

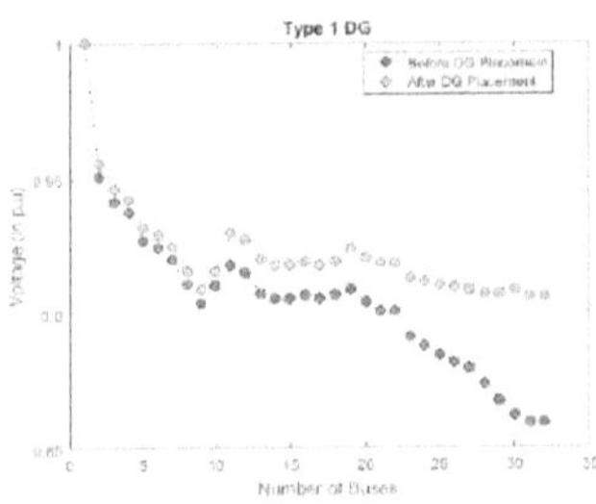
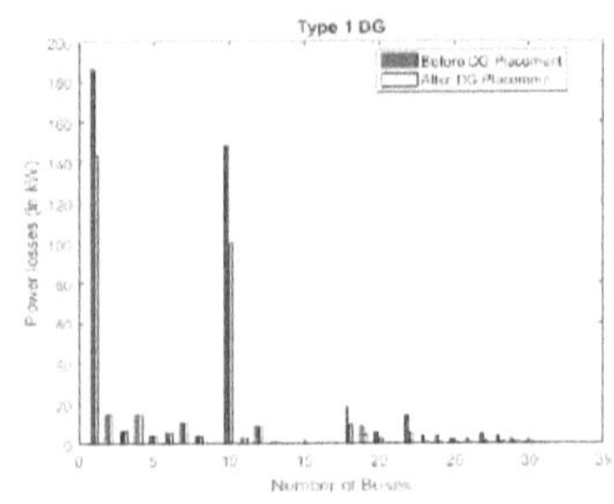

Fig 4.65: Voltage profile and Power losses of the practical test system for penetration of solar DG – scenario 2

Case-1, Scenario-3: 30 % integration of a SOLAR DG

The results from developed MSFA-PSO algorithm for allocation of a Solar DG unit for scenario-3 are tabulated in table 4.55. The system losses reduced to 306.12 kW from 477.99 kW and percentage losses are reduced to 35.95 %. Minimum voltage before DG placement is 0.85952 p.u at bus 32 and after DG placement the minimum voltage obtained to be 0.90936 p.u.at bus 32. Figures shows the variation of voltage and power losses at different buses before and after integration of 30% of Solar DG.

Table 4.55: Allocation of Solar DG for scenario 3

Particulars	MSFA 30% SOLAR DG	MSFA-PSO 30% SOLAR DG
1: OLDG	Bus 32	Bus 28
2: ODGS	1114.5 kW	1114.5 kW
3: BCPL (kW)	477.09	477.09
4: PLDG (kW)	300.51	306.12
5: PLR (%)	37.13	35.95
6: Min V w/o DG (p.u.)	Bus 32	Bus 32
	0.85952	0.85952
7: Min V W DG (p.u.)	Bus 09	Bus 09
	0.91342	0.90936

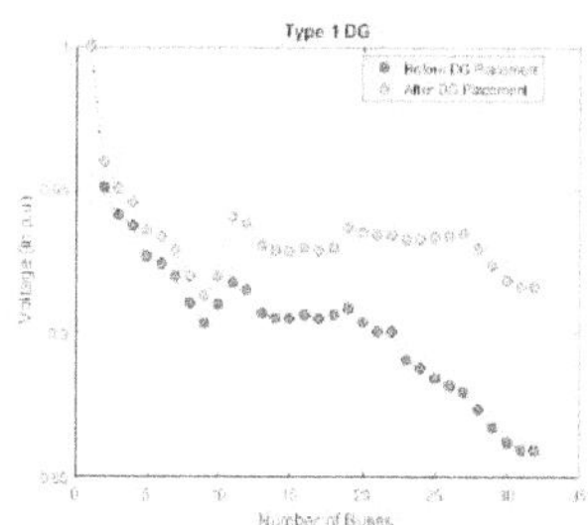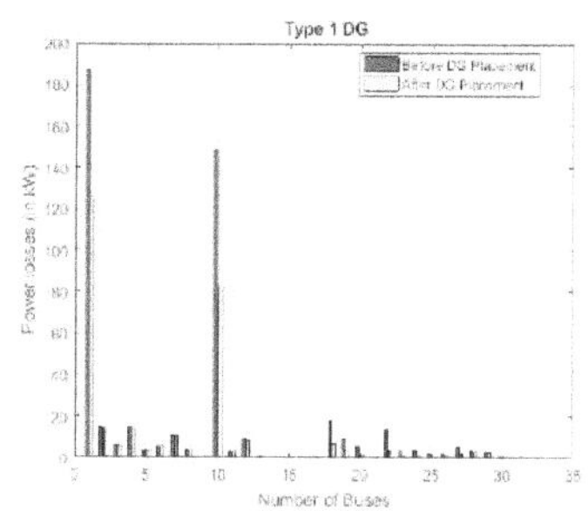

Fig 4.66: Voltage profile and Power losses of the practical test system for penetration of solar DG – scenario 3

Case-1, Scenario-4: 40 % integration of a SOLAR DG:

The results from developed MSFA-PSO algorithm for allocation of a Solar DG unit for scenario-4 are tabulated in table 4.56. The system losses reduced to 278.16 kW from 477.99 kW and percentage losses are reduced to 41.8 %. Minimum voltage before DG placement is 0.85952 p.u at bus 32 and after DG placement the minimum voltage obtained to be 0.91318 p.u.at bus 09. Figure 4.67 shows the variation of voltage and power losses at different buses before and after integration of 40% of Solar DG.

Table 4.56: Allocation of Solar DG for scenario 4

Particulars	MSFA 40% SOLAR DG	MSFA-PSO 40% SOLAR DG
1: OLDG	Bus 32	Bus 27
2: ODGS	1486 kW	1486 kW,
3: BCPL (kW)	477.09	477.09
4: PLDG (kW)	294.12	278.16
5: PLR (%)	38.35	41.8
6: Min V w/o DG (p.u.)	Bus 32	Bus 32
	0.85952	0.85952
7: Min V W DG (p.u.)	Bus 09	Bus 09
	0.91709	0.91318

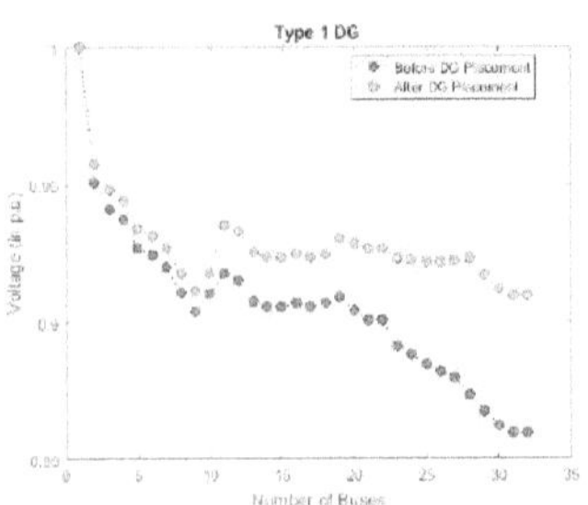
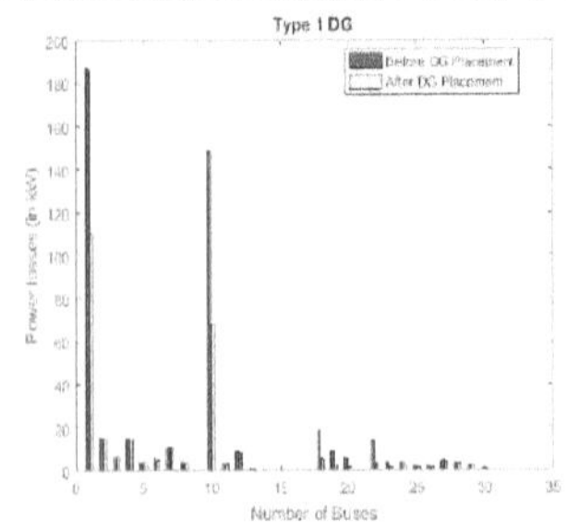

Fig 4.67: Voltage profile and Power losses of the practical test system for penetration of solar DG – scenario 4

Case-1, Scenario-5: 50 % integration of a SOLAR DG:

The results from developed MSFA-PSO algorithm for allocation of a Solar DG unit for scenario-5 are tabulated in table 4.57. The system losses reduced to 256.53 kW from 477.99 kW and percentage losses are reduced to 46.33 %. Minimum voltage before DG placement is 0.85952 p.u at bus 32 and after DG placement the minimum voltage obtained to be 0.91528 p.u.at bus 09.Figure 4.68 shows the variation of voltage and power losses at different buses before and after integration of 50% of Solar DG.

Table 4.57: Allocation of Solar DG for scenario 5

Particulars	MSFA 50% SOLAR DG	MSFA-PSO 50% SOLAR DG
1: OLDG	Bus 31	Bus 25
2: ODGS	1857.5 kW	1857.5 kW
3: BCPL (kW)	477.09	477.09
4: PLDG (kW)	268.2	256.53
5: PLR (%)	43.89	46.33
6: Min V w/o DG (p.u.)	Bus 32	Bus 32
	0.85952	0.85952
7: Min V W DG (p.u.)	Bus 09	Bus 09
	0.92577	0.91528

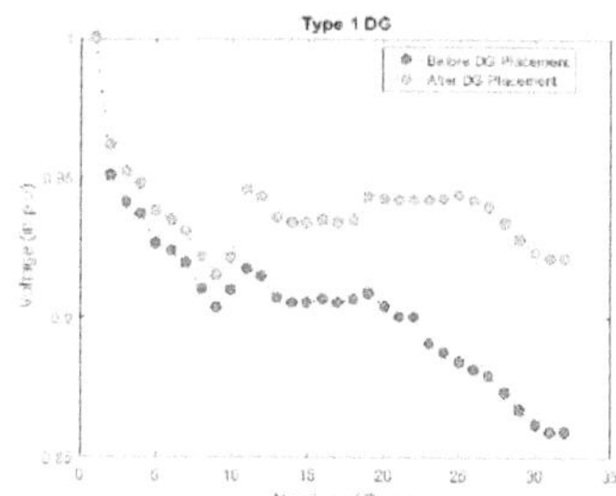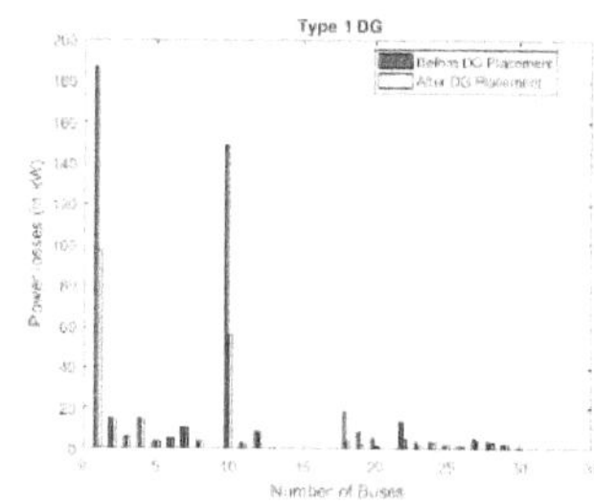

Fig 4.68: Voltage profile and Power losses of the practical test system for penetration of solar DG – scenario 5

Case-1, Scenario 1- : *optimal placement of a SOLAR DG*

The results obtained from the developed MSFA-PSO algorithm for case-1, scenario-1 and comparison with SFA is tabulated in table.4.1. Total losses of the system are reduced to 220.33 KW from 477.99KW with percentage losses are reduced to 53.90 %.

Table 4.58: Optimal allocation of Solar DG by MSFA-PSO and compared with MSFA

Particulars	MSFA SOLAR DG	MSFA-PSO SOLAR DG
1: OLDG	Bus 31	Bus 19
2: ODGS	2153.2 kW	2615.3 kW
3: BCPL (kW)	477.09	477.09
4: PLDG (kW)	225.65	220.33
5: PLR (%)	52.70	53.90
6: Min V w/o DG (p.u.)	Bus 32	Bus 32
	0.85952	0.85952
7: Min V W DG (p.u.)	Bus 09	Bus 09
	0.91982	0.92464

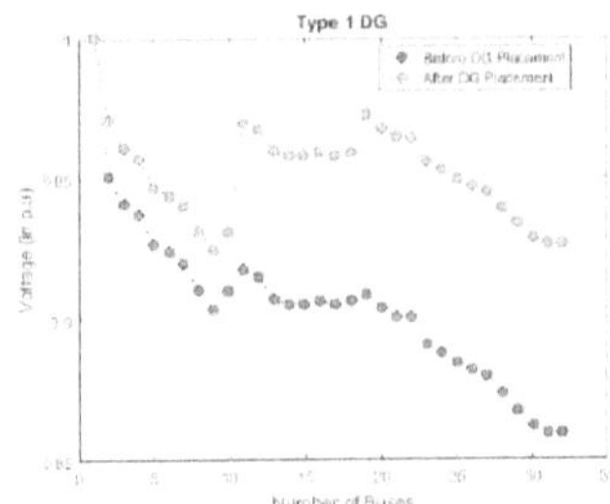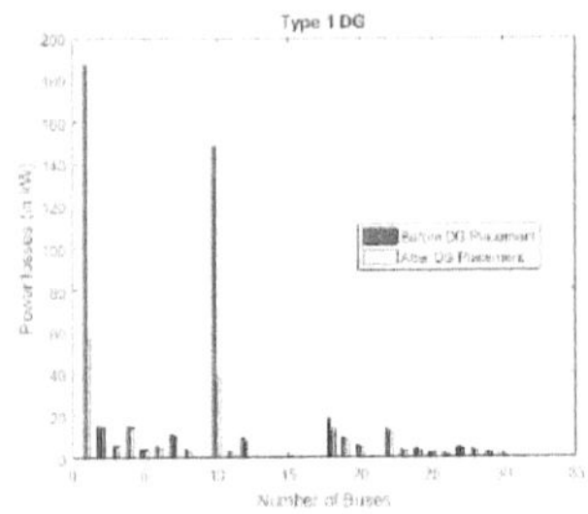

Fig 4.69: Voltage profile and Power losses of the practical test system for penetration of 1 solar DG – scenario 1

Case-1, Scenario-2: optimal placement of Two SOLAR DG units

The results obtained from the developed MSFA-PSO algorithm for case-1, scenario-2 is tabulated in table.4.59. Total losses of the system are reduced to 167.22 kW from 477.99 kW.

Table 4.59: Optimal allocation of two Solar DG by MSFA-PSO

Particulars	MSFA-PSO 2 SOLAR DG
1: OLDG	Bus 21, 08
2: ODGS	2684.4 kW, 1702.1 kW
3: BCPL (kW)	477.09
4: PLDG (kW)	167.22
5: PLR (%)	65.01
6: Min V w/o DG (p.u.)	Bus 32 0.85952
7: Min V W DG (p.u.)	Bus 32 0.92992

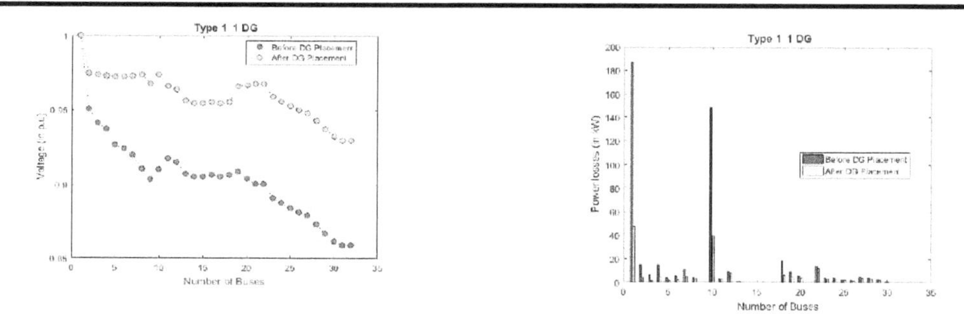

Fig 4.70: Voltage profile and Power losses of the practical test system for penetration of 2 solar DG – scenario 1

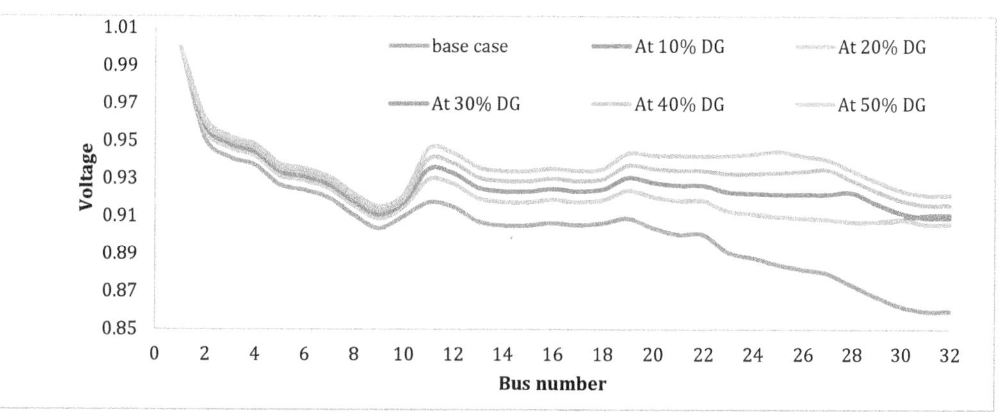

Fig 4.71: Voltage profile of the practical test system for penetration from 10% to 50%

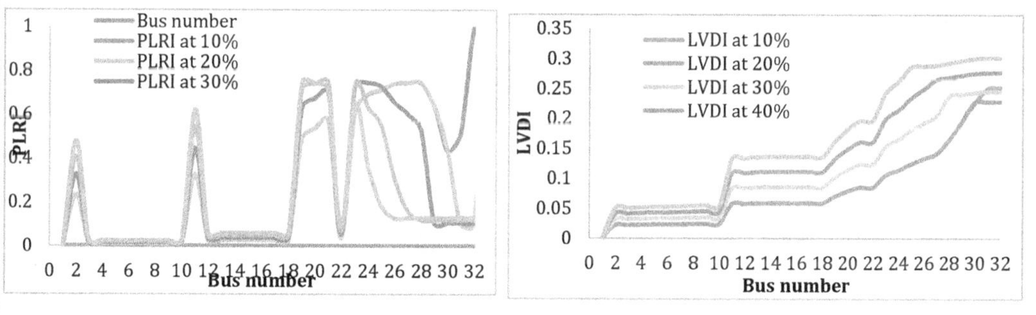

Fig 4.72: LVDI AND PLRI variation of the practical test system for penetration from 10% to 50%

Table 4.60: WSOF values at different buses after integration of wind DG at different penetration levels

Bus number	WSOF at 10%	WSOF at 20%	WSOF at 30%	WSOF at 40%	WSOF at 50%
1	0	0	0	0	0
2	0.106418	0.106638	0.149513	0.187592	0.221541
3	0.017693	0.017747	0.02561	0.033077	0.040368
4	0.017832	0.017867	0.025775	0.033307	0.040658
5	0.01813	0.018186	0.026239	0.033918	0.041362
6	0.018272	0.018311	0.026369	0.034092	0.041619
7	0.018395	0.018431	0.02655	0.034341	0.041918
8	0.018654	0.018711	0.026954	0.034881	0.042578
9	0.018947	0.018998	0.027379	0.035383	0.043163
10	0.018638	0.018708	0.026969	0.034872	0.04253
11	0.164637	0.164952	0.228141	0.281955	0.327546
12	0.04514	0.045222	0.065552	0.085085	0.104087
13	0.045711	0.045818	0.066416	0.086192	0.105428
14	0.045899	0.046002	0.066678	0.086518	0.105822
15	0.0459	0.046006	0.066679	0.086525	0.105827
16	0.045806	0.045904	0.066526	0.086337	0.105589
17	0.045901	0.046005	0.06666	0.086505	0.105812
18	0.045796	0.045883	0.066506	0.086306	0.105555
19	0.236441	0.236889	0.311998	0.364494	0.396315
20	0.26126	0.261767	0.337882	0.385002	0.405375
21	0.280279	0.280799	0.355654	0.395396	0.402605

22	0.066307	0.066454	0.096172	0.124683	0.152318
23	0.309319	0.309888	0.383997	0.414297	0.403882
24	0.34425	0.344827	0.396151	0.380338	0.301592
25	0.359386	0.35998	0.39904	0.359488	0.246001
26	0.375275	0.37583	0.377307	0.273479	0.223477
27	0.381982	0.382538	0.362798	0.223365	0.224494
28	0.394778	0.395314	0.351326	0.208385	0.226812
29	0.369907	0.370154	0.186819	0.210812	0.229453
30	0.309509	0.309406	0.188603	0.212823	0.231647
31	0.356539	0.176734	0.189548	0.213852	0.232747
32	0.547289	0.176723	0.18953	0.213866	0.232751

Case-2, Scenario-1: 10 % integration of a wind DG

The results from developed MSFA-PSO algorithm for allocation of a wind DG unit for scenario-1 are tabulated in table below. The analysis is performed for three different cases which is further compared to the previously calculated value of the modified shuffled frog leap algorithm. The first technique allocates the DG according to the fixed value of active and reactive power of the introduced wind generator in the system. Here, the system losses are reduced to 365.26 kW from 477.09 kW and percentage losses are reduced to 23.58. The next technique allocates the DG according to the fixed value of active power and optimal value of the reactive power which is calculated by the algorithm. Here, the system losses reduced are to 366.98 kW from 477.09 kW and percentage losses are reduced to 23.22. The last approach allocates the DG according to the optimal value of the active and reactive power which is calculated by the algorithm. Here, the system losses reduced are to 371.23 kW from 477.09 kW and percentage losses are reduced to 22.18. Minimum voltage before DG placement is 0.85952 p.u at bus 32 and after DG placement the minimum voltage obtained from the best of the three approaches is observed to be 0.92131 p.u.at bus 09.

Table 4.61: Allocation of wind DG for scenario 1

Particulars	MSFA 10% wind DG	MSFA-PSO 10% wind DG (fixed P, Q)	MSFA-PSO 10% wind DG (Fixed P, Optimal Q within 10%)	MSFA-PSO wind DG (Optimal P, Q within 10%)
1: OLDG	Bus 32	Bus 28	Bus 28	Bus 26
2: ODGS	371.5 kW -230.0 kVAR	371.5 kW -230.0 kVAR	371.5 kW -212.36 kVAR	312.23 kW -196.2 kVAR,
3: BCPL (kW)	477.09	477.09	477.09	477.09
4: PLDG (kW)	369.2	365.26	366.98	371.23
5: PLR (%)	22.46	23.58	23.22	22.18
6: Min V w/o DG (p.u.)	Bus 32 0.85952	Bus 32 0.85952	Bus 32 0.85952	Bus 32 0.85952
7: Min V W DG (p.u.)	Bus 09 0.91213	Bus 32 0.92131	Bus 09 0.89212	Bus 09 0.9123

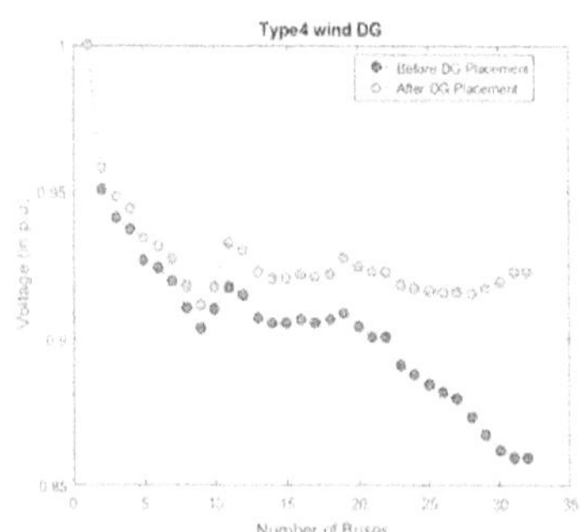

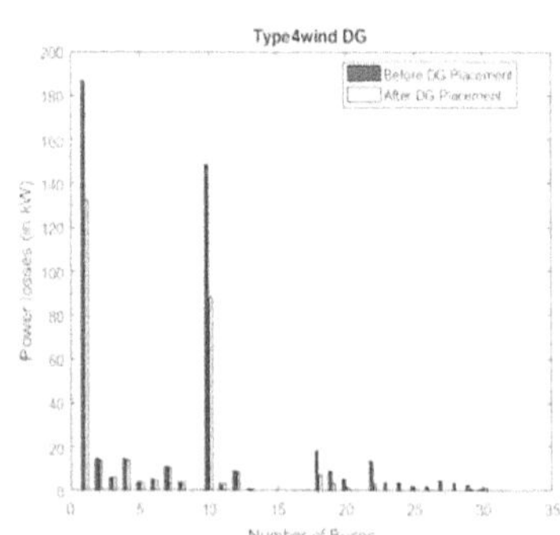

Fig 4.73: Voltage profile and Power losses of the practical test system for penetration of wind DG – scenario 1

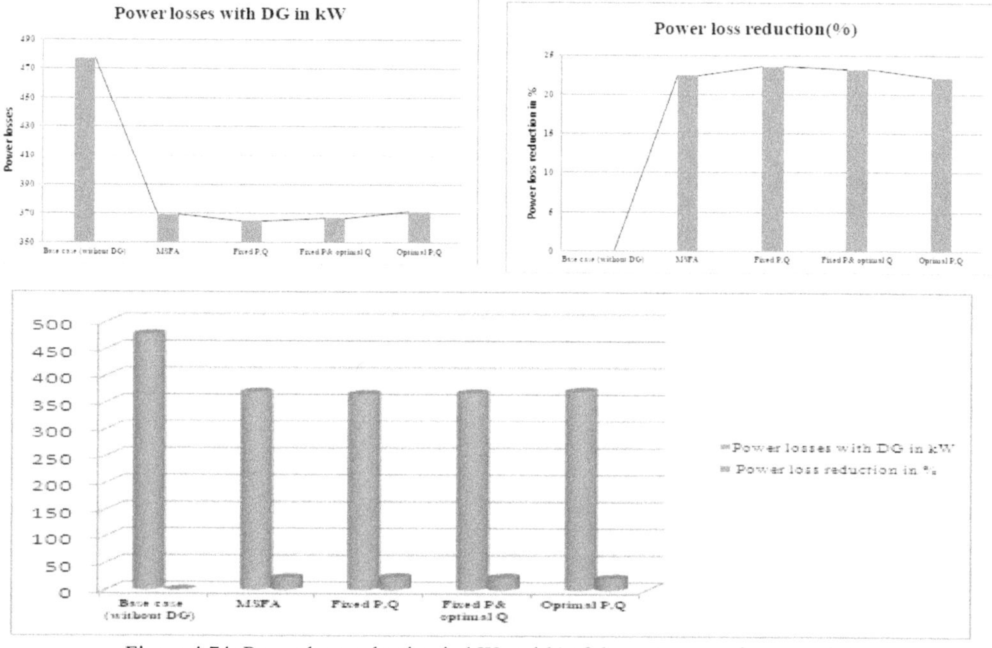

Figure 4.74: Power loss reduction in kW and % of the test system for scenario 1

Case-2, Scenario-2: 20 % integration of a wind DG

The results from developed MSFA-PSO algorithm for allocation of a wind DG unit for scenario-1 are tabulated in table below. The analysis is performed for three different cases which is further compared to the previously calculated value of the modified shuffled frog leap algorithm. The first technique allocates the DG according to the fixed value of active and reactive power of the introduced wind generator in the system. Here, the system losses are reduced to 296.43 kW from 477.09 kW and percentage losses are reduced to 37.98. The next technique allocates the DG according to the fixed value of active power and optimal value of the reactive power which is calculated by the algorithm. Here, the system losses reduced are to 310.4 kW from 477.09 kW and percentage losses are reduced to 35.4. The last approach allocates the DG according to the optimal value of the active and reactive power which is calculated by the algorithm. Here, the system losses reduced are to 302.55 kW from 477.09 kW and percentage losses are reduced to 36.7. Minimum voltage before DG placement is 0.85952 p.u at bus 32 and after DG placement the minimum voltage obtained from the best of the three approaches is observed to be 0.92149 p.u.at bus .Figures shows the graphs of voltage and power losses at different buses before and after integration of 20% of wind DG.

Table 4.62: Allocation of wind DG for scenario 2

Particulars	MSFA 20% wind DG	MSFA-PSO 20% wind DG (fixed P, Q)	MSFA-PSO 20% wind DG (Fixed P, Optimal Q within 20%)	MSFA-PSO wind DG (Optimal P, Q within 20%)
1: OLDG	Bus 31	Bus 25	Bus 32	Bus 32
2: ODGS	743 kW 460 kVAR	743 kW -460 kVAR	743 kW -386 kVAR	743 kW 460 kVAR
3: BCPL (kW)	477.09	477.09	477.09	477.09
4: PLDG (kW)	298.31	296.43	310.4	302.55
5: PLR (%)	37.47	37.98	35.4	36.7
6: Min V w/o DG (p.u.)	Bus 32 0.85952	Bus 32 0.85952	Bus 32 0.85952	Bus 32 0.85952
7: Min V W DG (p.u.)	Bus 09 0.92810	Bus 32 0.91091	Bus 09 0.91039	Bus 09 0.92282

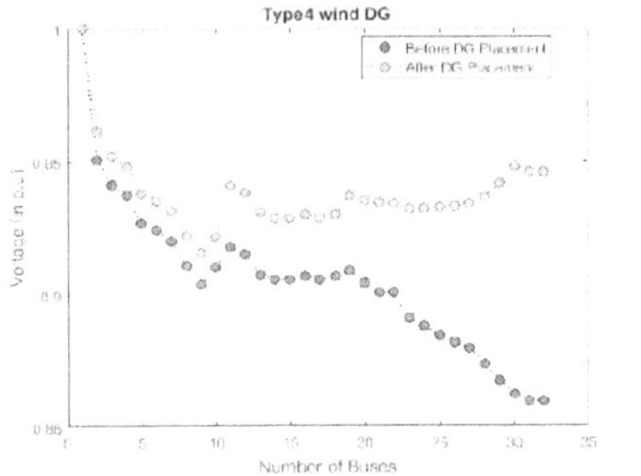

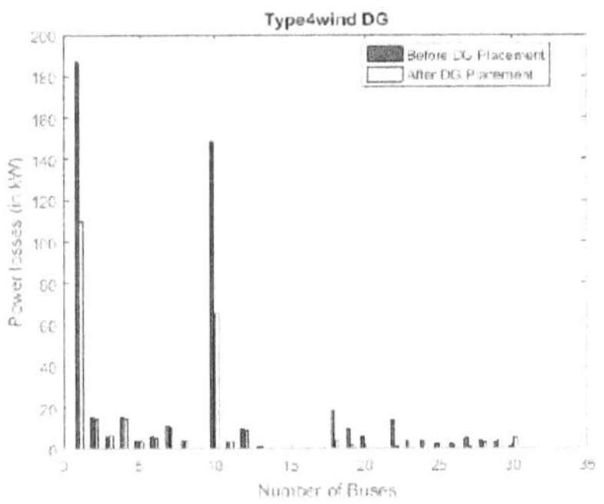

Fig 4.75: Voltage profile and Power losses of

the practical test system for penetration of wind DG – scenario 2

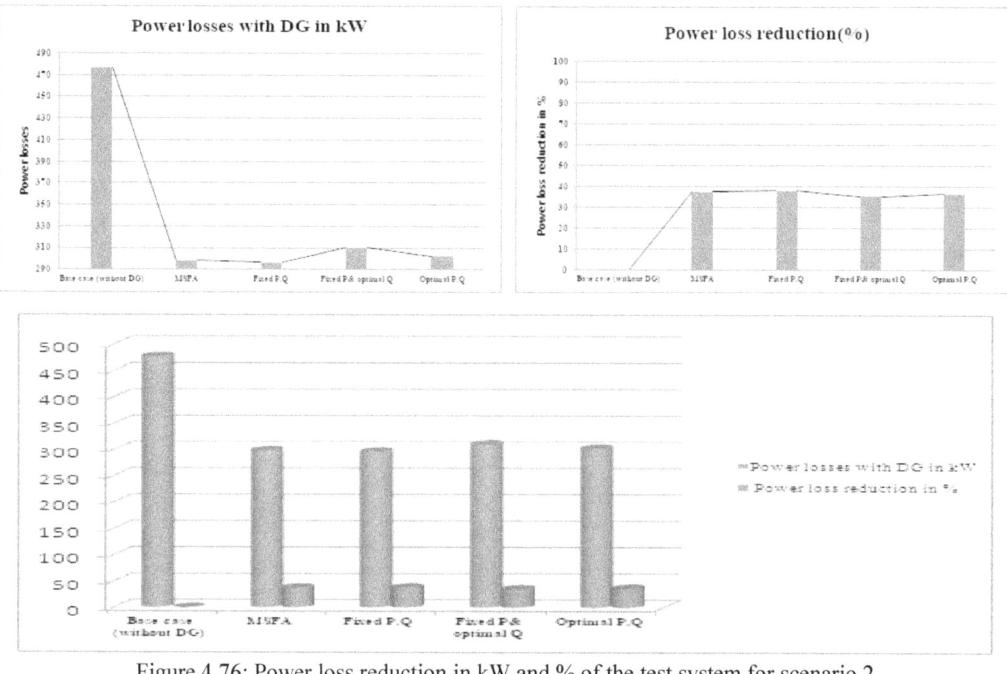

Figure 4.76: Power loss reduction in kW and % of the test system for scenario 2

Case-2, Scenario-3: 30 % integration of a wind DG

The results from developed MSFA-PSO algorithm for allocation of a wind DG unit for scenario-1 are tabulated in table below. The analysis is performed for three different cases which is further compared to the previously calculated value of the modified shuffled frog leap algorithm. The first technique allocates the DG according to the fixed value of active and reactive power of the introduced wind generator in the system. Here, the system losses are reduced to 248.82 kW from 477.09 kW and percentage losses are reduced to 47.94. The next technique allocates the DG according to the fixed value of active power and optimal value of the reactive power which is calculated by the algorithm. Here, the system losses reduced are to 259.59 kW from 477.09 kW and percentage losses are reduced to 45.69. The last approach allocates the DG according to the optimal value of the active and reactive power which is calculated by the algorithm. Here, the system losses reduced are to 244.24 kW from 477.09 kW and percentage losses are reduced to 48.9. Minimum voltage before DG placement is 0.85952 p.u at bus 32 and after DG placement the minimum voltage obtained from the best of the three approaches is observed to be 0.92698 p.u.at bus 09.

Table 4.63: Allocation of wind DG for scenario 3

Particulars	MSFA 30% wind DG	MSFA-PSO 30% wind DG (fixed P,Q)	MSFA-PSO 30% wind DG (Fixed P, Optimal Q within 30%)	MSFA-PSO wind DG (Optimal P , Q within 30%)
1: OLDG	Bus 31	Bus 31	Bus 30	Bus 31
2: ODGS	1145 kW -690 kVAR	1145 kW -690 kVAR	1145 kW -458 kVAR	948.23 kW -585.38 kVAR
3: BCPL (kW)	477.09	477.09	477.09	477.09
4: PLDG (kW)	261.98	248.82	259.59	244.24
5: PLR (%)	45.19	47.94	45.69	48.9
6: Min V w/o DG (p.u.)	Bus 32 0.85952	Bus 32 0.85952	Bus 32 0.85952	Bus 32 0.85952
7: Min V W DG (p.u.)	Bus 09 0.92822	Bus 09 0.92698	Bus 09 0.92518	Bus 09 0.92627

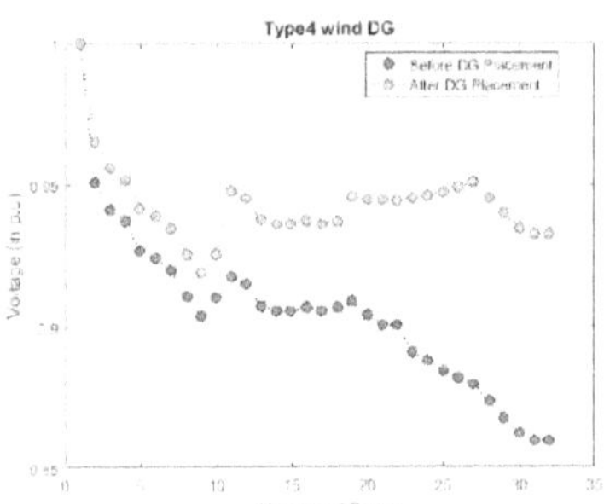
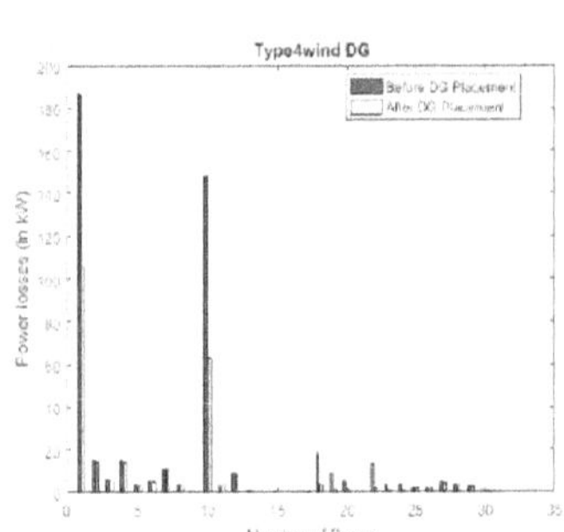

Fig 4.77: Voltage profile and Power losses of the practical test system for penetration of wind DG – scenario 3

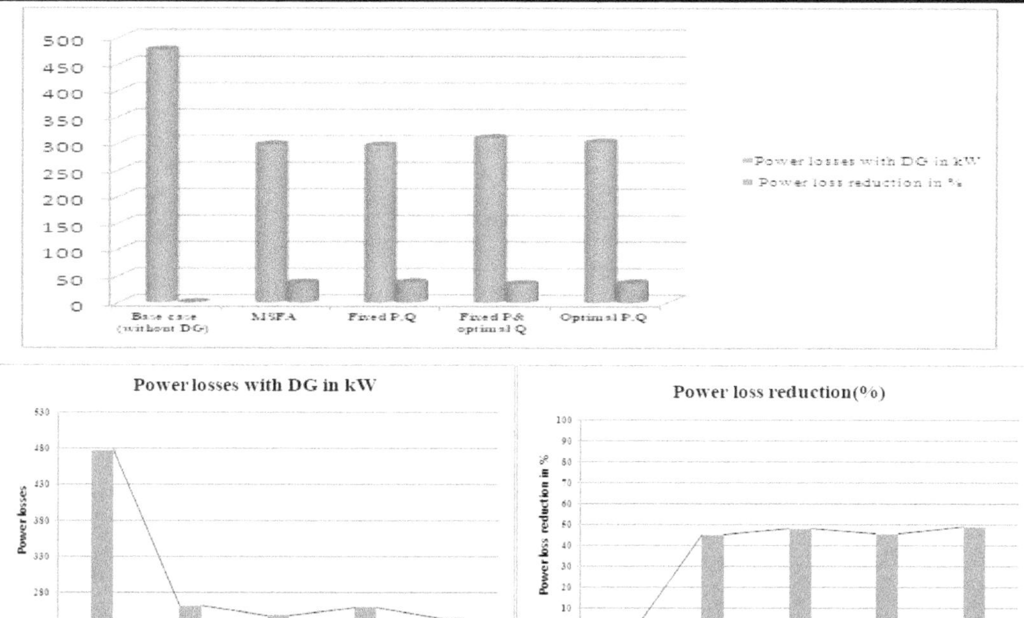

Figure 4.78: Power loss reduction in kW and % of the test system for scenario 3

Case-2, Scenario-4: 40 % integration of a wind DG

The results from developed MSFA-PSO algorithm for allocation of a wind DG unit for scenario-1 are tabulated in table below. The analysis is performed for three different cases which is further compared to the previously calculated value of the modified shuffled frog leap algorithm. The first technique allocates the DG according to the fixed value of active and reactive power of the introduced wind generator in the system. Here, the system losses are reduced to 206.99 kW from 477.09 kW and percentage losses are reduced to 56.69. The next technique allocates the DG according to the fixed value of active power and optimal value of the reactive power which is calculated by the algorithm. Here, the system losses reduced are to 218.68 kW from 477.09 kW and percentage losses are reduced to 54.68. The last approach allocates the DG according to the optimal value of the active and reactive power which is calculated by the algorithm. Here, the system losses reduced are to 204.73 kW from 477.09 kW and percentage losses are reduced to 57.16. Minimum voltage before DG placement is 0.85952 p.u at bus 32 and after DG placement the minimum voltage obtained from the best of the three approaches is observed to be 0.92783 p.u.at bus 09. Figures shows the graphs of voltage and power losses at different buses before and after integration of 40% of wind DG.

Table 4.64: Allocation of wind DG for scenario 4

Particulars	MSFA 40% wind DG	MSFA-PSO 40% wind DG (fixed P,Q)	MSFA-PSO 40% wind DG (Fixed P, Optimal Q within 40%)	MSFA-PSO wind DG (Optimal P , Q within 40%)
1: OLDG	Bus 31	Bus 28	Bus 27	Bus 27
2: ODGS	1486 kW 920 kVAR	1486 kW -920 kVAR	1486 kW -651.13 kVAR	1422.3 kW -769.23 kVAR
3: BCPL (kW)	477.09	477.09	477.09	477.09
4: PLDG (kW)	245.12	206.99	218.68	204.73
5: PLR (%)	48.62	56.69	54.68	57.16
6: Min V w/o DG (p.u.)	Bus 32 0.85952	Bus 32 0.85952	Bus 32 0.85952	Bus 32 0.85952
7: Min V W DG (p.u.)	Bus 09 0.9251	Bus 09 0.92783	Bus 09 0.91906	Bus 32 0.92099

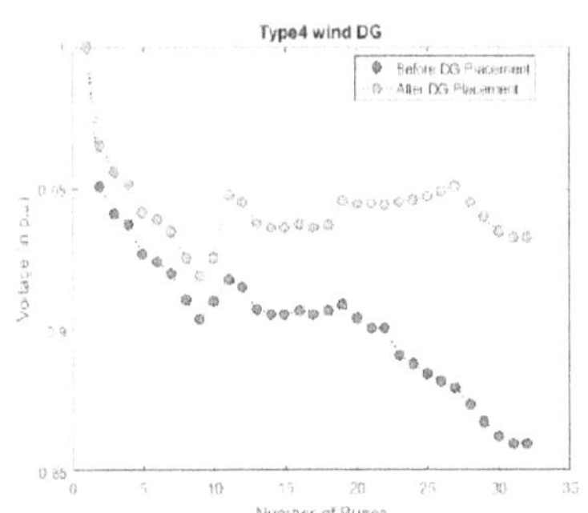

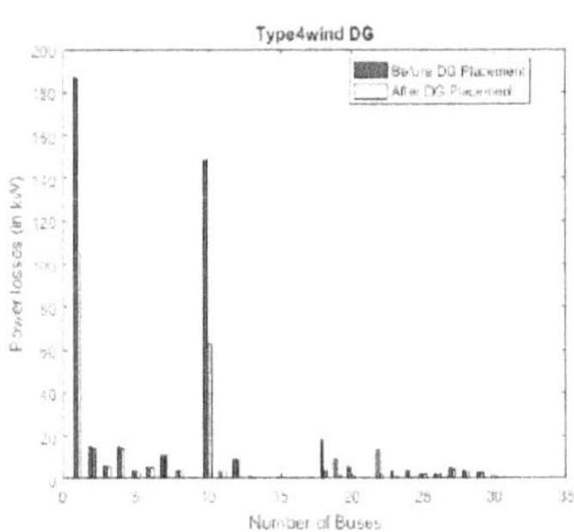

Fig 4.79: Voltage profile and Power losses of the practical test system for penetration of wind DG – scenario 4

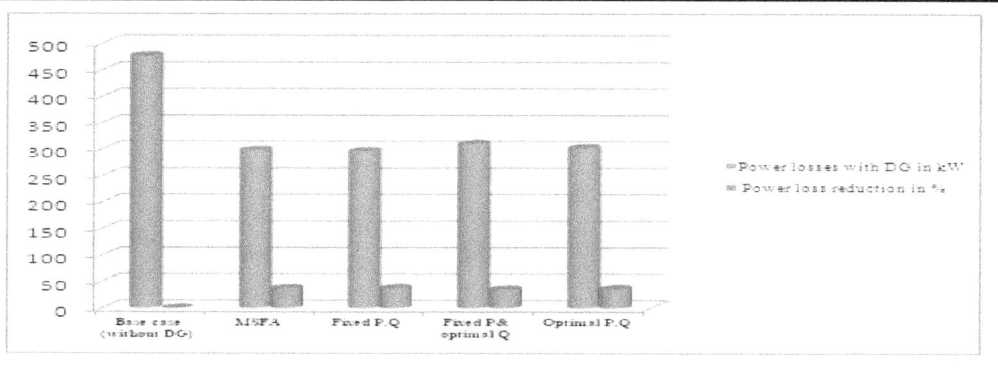

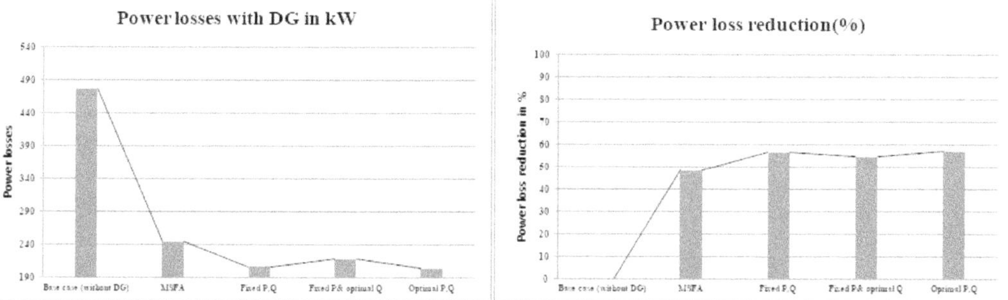

Figure 4.80: Power loss reduction in kW and % of the test system for scenario 4

Case-2, Scenario-5: 50 % integration of a wind DG

The results from developed MSFA-PSO algorithm for allocation of a wind DG unit for scenario-1 are tabulated in table below. The analysis is performed for three different cases which is further compared to the previously calculated value of the modified shuffled frog leap algorithm. The first technique allocates the DG according to the fixed value of active and reactive power of the introduced wind generator in the system. Here, the system losses are reduced to 202.12 kW from 477.09 kW and percentage losses are reduced to 57.63. The next technique allocates the DG according to the fixed value of active power and optimal value of the reactive power which is calculated by the algorithm. Here, the system losses reduced are to 207.85 kW from 477.09 kW and percentage losses are reduced to 56.51. The last approach allocates the DG according to the optimal value of the active and reactive power which is calculated by the algorithm. Here, the system losses reduced are to 199.99 kW from 477.09 kW and percentage losses are reduced to 58.08. Minimum voltage before DG placement is 0.85952 p.u at bus 32 and after DG placement the minimum voltage obtained from the best of the three approaches is observed to be 0.93393 p.u.at bus 32. Figures shows the graphs of voltage and power losses at different buses before and after integration of 50% of wind DG.

Table 4.65: Allocation of wind DG for scenario 5

Particulars	MSFA 50% wind DG	MSFA-PSO 50% wind DG (fixed P, Q)	MSFA-PSO 50% wind DG (Fixed P, Optimal Q within 50%)	MSFA-PSO wind DG (Optimal P, Q within 50%)
1: OLDG	Bus 22	Bus 25	Bus 25	Bus 26
2: ODGS	1857.5 kW 1094.1 kVAR	1857.5 kW -1094.1 kVAR	1857.5 kW -462.15 kVAR	1740.6 kW -825.7 kVAR
3: BCPL (kW)	477.09	477.09	477.09	477.09
4: PLDG (kW)	238.11	202.12	207.85	199.99
5: PLR (%)	50.18	57.63	56.51	58.08
6: Min V w/o DG (p.u.)	Bus 32 0.85952	Bus 32 0.85952	Bus 32 0.85952	Bus 32 0.85952
7: Min V W DG (p.u.)	Bus 09 0.92277	Bus 09 0.92492	Bus 09 0.91999	Bus 32 0.93393

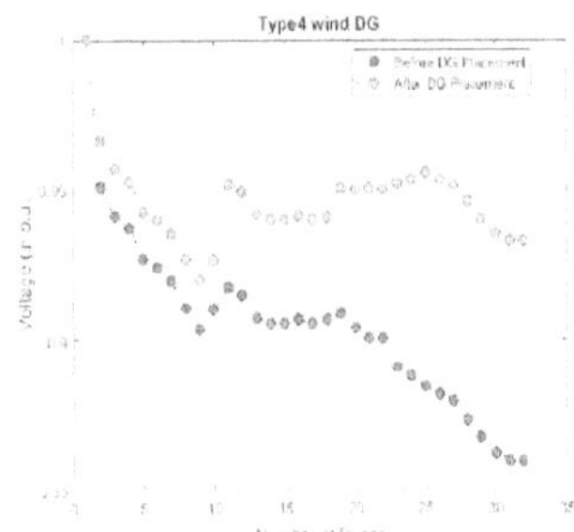

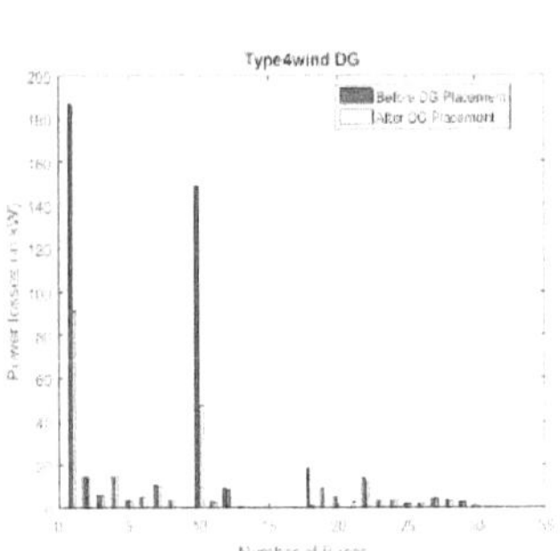

Fig 4.81: Voltage profile and Power losses of practical test system for penetration of wind DG – scenario 5

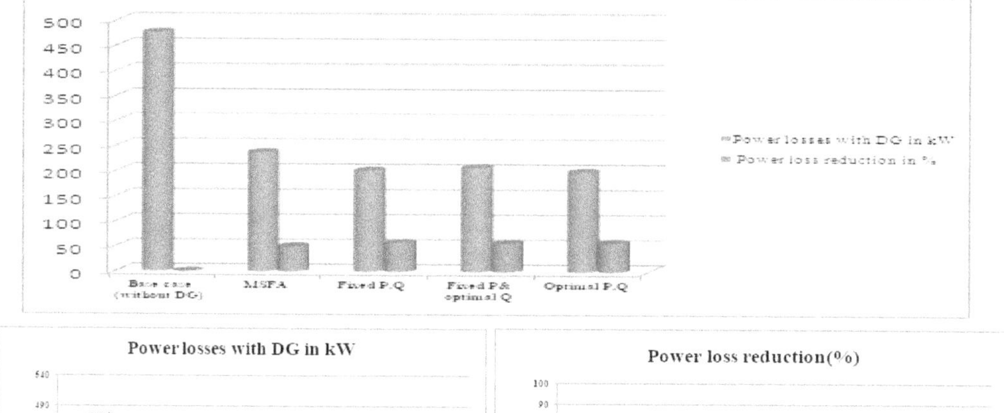

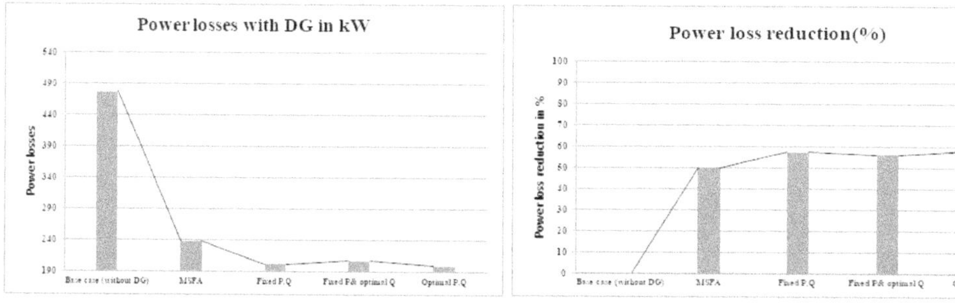

Figure 4.82: Power loss reduction in kW and % of the test system for scenario 5

Case-2, Scenario 1- : optimal placement of a WIND DG

The results obtained from the developed MSFA-PSO algorithm for case-2, scenario-1 and comparison with SFA is tabulated in table.4.3. Total losses of the system are reduced to 141.55 kW from 477.99 kW with percentage losses are reduced to 70.38%.

Table 4.66: Optimal allocation of Wind DG by MSFA-PSO and compared with MSFA

Particulars	MSFA WIND DG	MSFA-PSO WIND DG
1: OLDG	Bus 31	Bus 21
2: ODGS	1355.9 kW, -613 kVAR	2732 kW, -1840 kVAR
3: BCPL (kW)	477.09	477.09
4: PLDG (kW)	145.32	141.55
5: PLR (%)	69.54	70.38
6: Min V w/o DG (p.u.)	Bus 32	Bus 32
	0.85952	0.85952

| 7: Min V W DG (p.u.) | Bus 09 | Bus 09 |
| | 0.9295 | 0.9351 |

Case-2, Scenario-2: optimal placement of Two WIND DG units

The results obtained from the developed MSFA-PSO algorithm for case-2, scenario-2 is tabulated in table.4.2. Total losses of the system are reduced to 118.85 KW from 477.99KW.

Table 4.67: Optimal allocation of two wind DG by MSFA-PSO

Particulars	MSFA-PSO, 2 WIND DGs
1: OLDG	Bus 19
	Bus 30
2: ODGS	1545 kW, -1206 kVAR
	1206 kW, -904 kVAR
3: BCPL (kW)	477.09
4: PLDG (kW)	118.85
5: PLR (%)	75.13
6: Min V w/o DG (p.u.)	Bus 32
	0.85952
7: Min V W DG (p.u.)	Bus 09
	0.9373

Table 4.68: WSOF values at different buses after integration of wind DG at different penetration levels

Bus number	WSOF at 10%	WSOF at 20%	WSOF at 30%	WSOF at 40%	WSOF at 50%
1	0	0	0	0	0
2	0.082248	0.151926	0.21314	0.267227	0.315411
3	0.016623	0.031819	0.046258	0.060053	0.073508
4	0.016738	0.032039	0.046577	0.060493	0.074009

5	0.017018	0.032588	0.047387	0.06156	0.075331
6	0.017158	0.03281	0.047663	0.061904	0.075743
7	0.017251	0.033052	0.048013	0.062341	0.07629
8	0.01754	0.033527	0.048734	0.06331	0.077474
9	0.017794	0.034036	0.049441	0.064192	0.078546
10	0.017522	0.033512	0.048725	0.063267	0.077419
11	0.124194	0.225864	0.311713	0.384021	0.444838
12	0.033207	0.063358	0.091773	0.118862	0.145009
13	0.033639	0.064162	0.092951	0.120386	0.1469
14	0.033771	0.064416	0.093321	0.120836	0.147432
15	0.033771	0.064419	0.093323	0.120845	0.147439
16	0.033716	0.064291	0.093104	0.120562	0.14713
17	0.033759	0.0644	0.093303	0.120833	0.147419
18	0.033706	0.064256	0.093094	0.120565	0.147084
19	0.185456	0.32264	0.423631	0.493165	0.534719
20	0.208156	0.355988	0.457981	0.51998	0.545994
21	0.226506	0.381445	0.481458	0.533344	0.541693
22	0.048556	0.092315	0.133394	0.172322	0.209777
23	0.254041	0.420502	0.519095	0.558105	0.54284
24	0.297735	0.466942	0.53443	0.51203	0.406709
25	0.317849	0.487112	0.537958	0.484025	0.333104
26	0.351026	0.507926	0.508257	0.369528	0.303936
27	0.36768	0.516723	0.488672	0.303083	0.305242
28	0.391552	0.533776	0.473191	0.283665	0.308353
29	0.453888	0.499779	0.254736	0.286913	0.311854
30	0.490073	0.419109	0.257144	0.289581	0.314738
31	0.466136	0.439148	0.258371	0.290965	0.316234

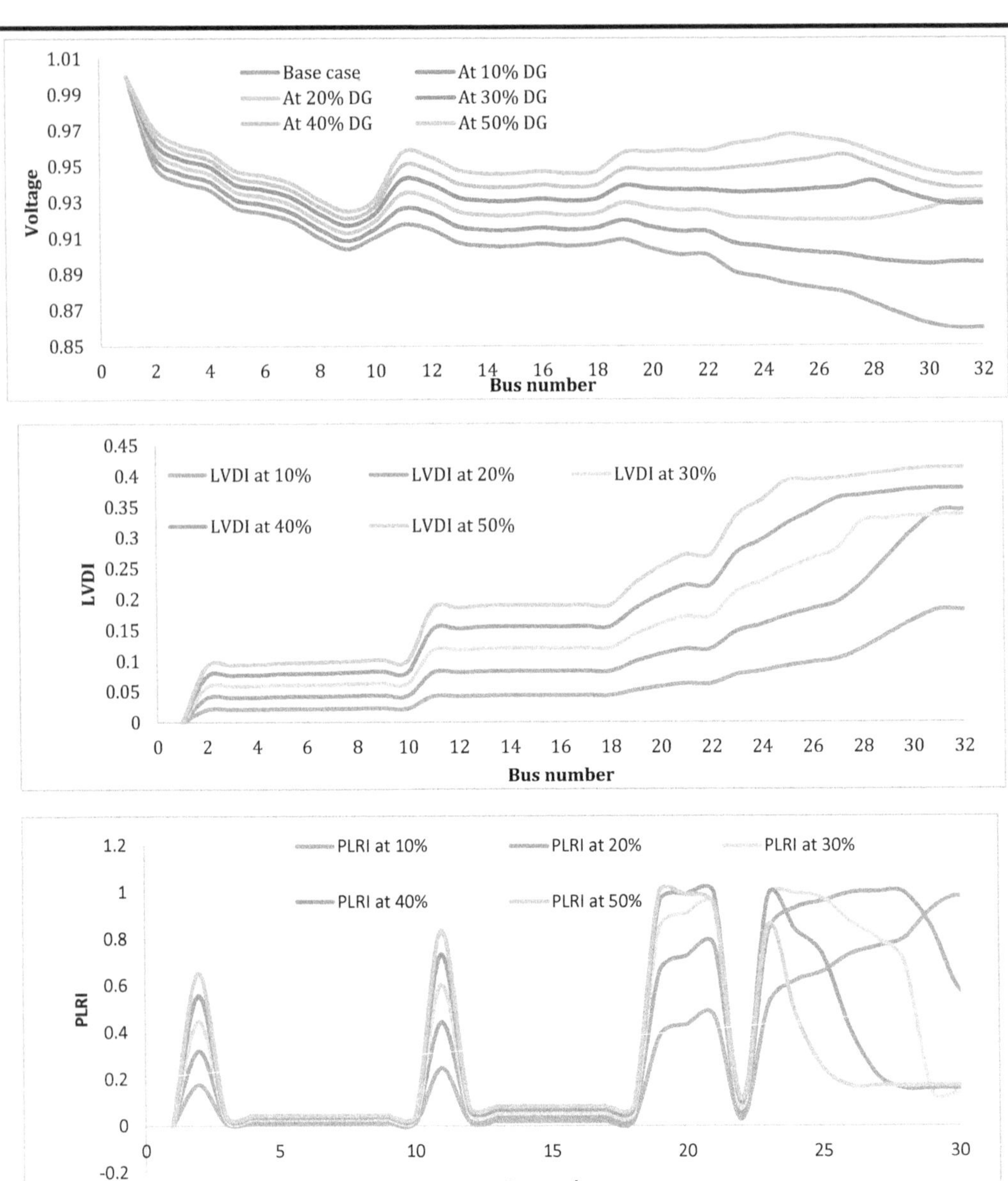

Figure 4.83: Voltage profile, LVDI AND PLRI variation of the practical test system for penetration from 10% to 50%

Case-3, Scenario-1: 10 % integration of SOLAR, WIND Hybrid DG

The results from developed MSFA-PSO algorithm for allocation of Hybrid DG unit for scenario-1 are tabulated in table 4. The system losses reduced to 283.7 kW from 477.99 kW and percentage losses are reduced to 40.648 %. Minimum voltage before DG placement is

0.85952 p.u at bus 32 and after DG placement the minimum voltage obtained to be 0.90769 p.u.at bus 32. Figures 4.84 shows the variation of voltage and power losses at different buses before and after integration of 10% of hybrid DG.

Table 4.69: Allocation of Hybrid DG for scenario 1

Particulars	MSFA Hybrid DG	MSFA-PSO Hybrid DG
1: OLDG	Bus 2 Bus 30	Bus 28 Bus 23
2: ODGS	743 KW, 371.5 KW, -220 KVAR	743 KW, 371.5 KW, -220 KVAR
3: BCPL (kW)	477.09	477.09
4: PLDG (kW)	286.86	283.7
5: PLR (%)	39.87	40.64
6: Min V w/o DG (p.u.)	Bus 32 0.85952	Bus 32 0.85952
7: Min V W DG (p.u.)	Bus 32 0.90231	Bus 32 0.90769

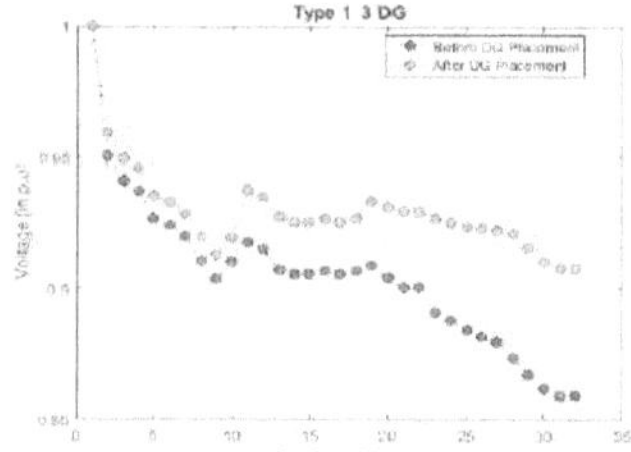
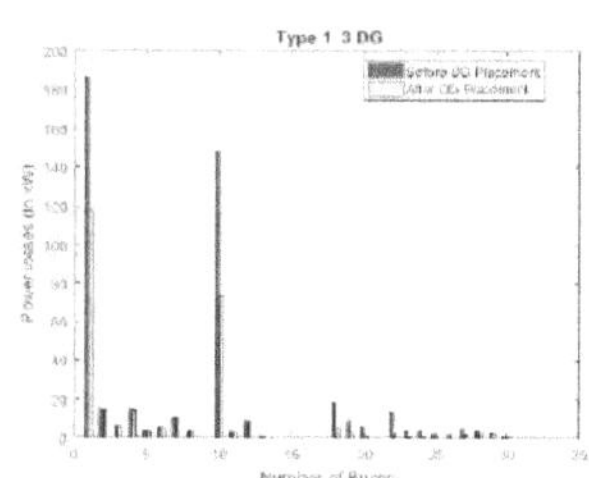

Figure 4.84: Voltage profile and Power losses of the system for penetration of Hybrid DG-scenario 1

Case-3, Scenario-2: 20 % integration of SOLAR, WIND Hybrid DG

The results from developed MSFA-PSO algorithm for allocation of Hybrid DG unit for scenario-2 are tabulated in table 4.70. The system losses reduced to 222.27 kW from 477.99

kW and percentage losses are reduced to 53.5 %. Minimum voltage before DG placement is 0.85952 p.u at bus 32 and after DG placement the minimum voltage obtained to be 0.91729 p.u.at bus 09. Figures 4.85 shows the variation of voltage and power losses at different buses before and after integration of 20% of hybrid DG.

Table 4.70: Allocation of Hybrid DG for scenario 2

Particulars	MSFA Hybrid DG	MSFA-PSO Hybrid DG
1: OLDG	Bus 06, 32	Bus 13,30
2: ODGS	743 kW, 743 kW, -460 kVAR	743 kW, 743 kW, -460 kVAR
3: BCPL (kW)	477.09	477.09
4: PLDG (kW)	276.76	222.27
5: PLR (%)	41.98	53.5
6: Min V w/o DG (p.u.)	Bus 32 0.85952	Bus 32 0.85952
7: Min V W DG (p.u.)	Bus 09 0.91321	Bus 09 0.91729

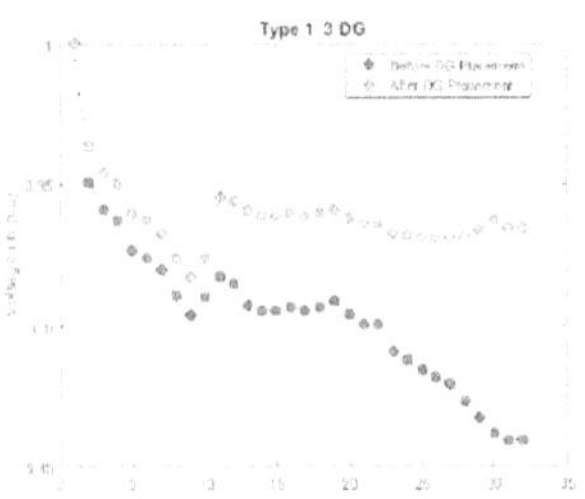
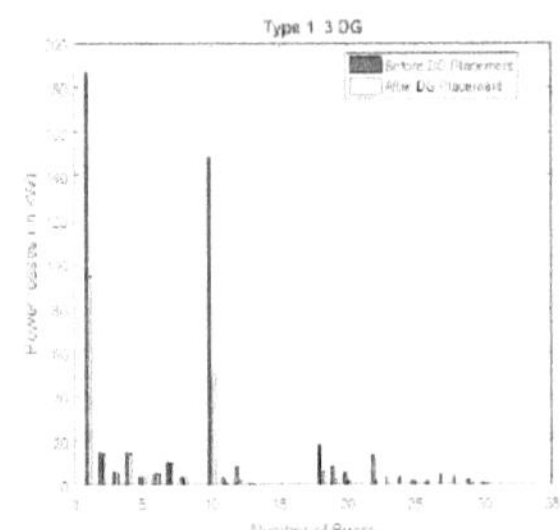

Figure 4.85: Voltage profile and Power losses of the system for penetration of Hybrid DG-scenario 2

Case-3, Scenario-3: 30 % integration of SOLAR, WIND Hybrid DG

The results from developed MSFA-PSO algorithm for allocation of Hybrid DG unit for scenario-3 are tabulated in table 4.72 The system losses reduced to 175.55 kW from 477.99 kW and percentage losses are reduced to 63.27 %. Minimum voltage before DG placement is 0.85952 p.u at bus 32 and after DG placement the minimum voltage obtained to be 0.92314 p.u.at bus 09. Figure 4.86 shows the variation of voltage and power losses at different buses before and after integration of 30% of hybrid DG.

Table 4.72: Allocation of Hybrid DG for scenario 3

Particulars	MSFA Hybrid DG	MSFA-PSO Hybrid DG
1: OLDG	Bus 02 Bus 30	Bus 29 Bus 16
2: ODGS	1114.5 kW, 1114.5 kW, -672.02 kVAR	1114.5 kW, 1114.5 kW, -672.02 kVAR
3: BCPL (kW)	477.09	477.09
4: PLDG (kW)	256.84	175.55
5: PLR (%)	46.26	63.27
6: Min V w/o DG (p.u.)	Bus 32 0.85952	Bus 32 0.85952
7: Min V W DG (p.u.)	Bus 09 0.92114	Bus 09 0.92314

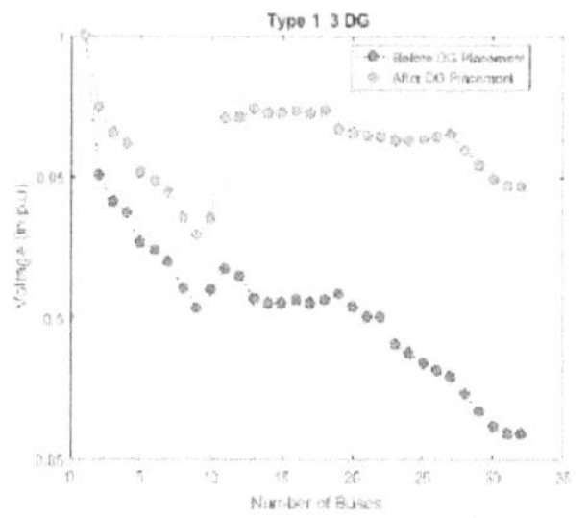
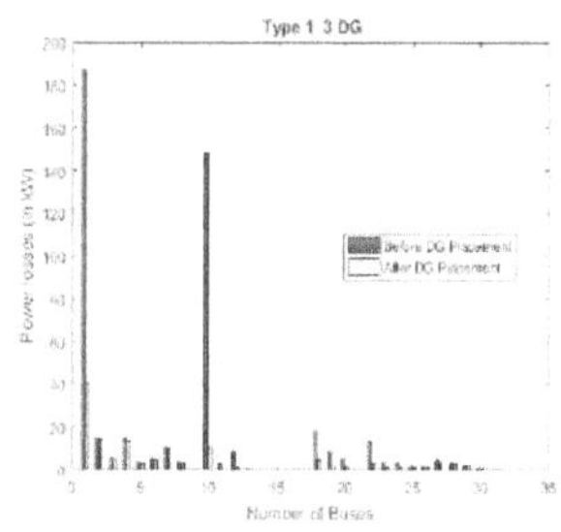

Figure 4.86: Voltage profile and Power losses of the system for penetration of Hybrid DG-scenario 3

Case-3, Scenario-4: 40 % integration of SOLAR, WIND Hybrid DG

The results from developed MSFA-PSO algorithm for allocation of Hybrid DG unit for scenario-4 are tabulated in table 4.73 The system losses reduced to 138.28 kW from 477.99 kW and percentage losses are reduced to 71.06 %. Minimum voltage before DG placement is 0.85952 p.u at bus 32 and after DG placement the minimum voltage obtained to be 0.92919 p.u.at bus 09. Figure 4.87 shows the variation of voltage and power losses at different buses before and after integration of 40% of hybrid DG.

Table 4.73: Allocation of Hybrid DG for scenario 4

Particulars	MSFA Hybrid DG	MSFA-PSO Hybrid DG
1: OLDG	Bus 27 Bus 30	Bus 27 Bus 13
2: ODGS	1486 kW, 1486 kW, -920 kVAR	1486 kW, 1486 kW, -920 kVAR
3: BCPL (kW)	477.09	477.09
4: PLDG (kW)	223.32	138.28
5: PLR (%)	53.19	71.06
6: Min V w/o DG (p.u.)	Bus 32 0.85952	Bus 32 0.85952
7: Min V W DG (p.u.)	Bus 09 0.92163	Bus 09 0.92919

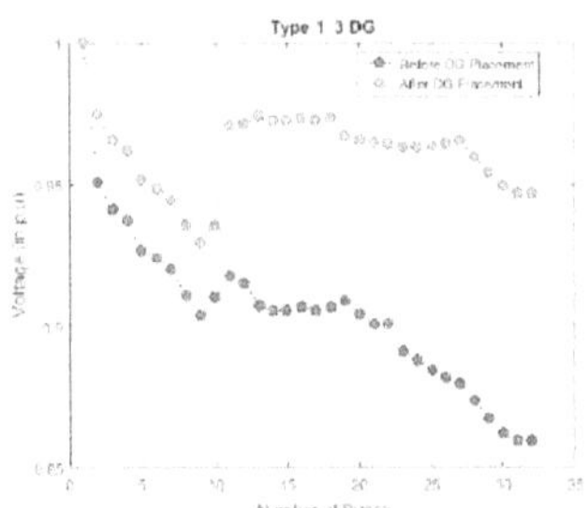
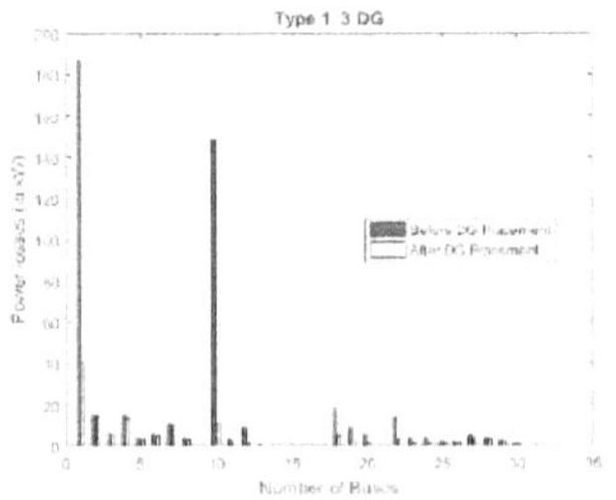

Figure 4.87: Voltage profile and Power losses of the system for penetration of Hybrid DG-scenario 4

Case-3, Scenario-5: 50 % integration of SOLAR, WIND Hybrid DG

The results from developed MSFA-PSO algorithm for allocation of Hybrid DG unit for scenario-5 are tabulated in table 4.74 The system losses reduced to 178.47 kW from 477.99 kW and percentage losses are reduced to 62.66 %. Minimum voltage before DG placement is 0.85952 p.u at bus 32 and after DG placement the minimum voltage obtained to be 0.9242 p.u.at bus 09. Figures shows the variation of voltage and power losses at different buses before and after integration of 50% of hybrid DG.

Table 4.74: Allocation of Hybrid DG for scenario 5

Particulars	MSFA Hybrid DG	MSFA-PSO Hybrid DG
1: OLDG	Bus 02, 30	Bus 28, 06
2: ODGS	1857.5 kW, 1857.5 kW, -1115 kVAR	1857.5 kW, 1857.5 kW, -1115 kVAR
3: BCPL (kW)	477.09	477.09
4: PLDG (kW)	212.94	151.98
5: PLR (%)	54.55	68.20
6: Min V w/o DG (p.u.)	Bus 32	Bus 32
	0.85952	0.85952
7: Min V W DG (p.u.)	Bus 09	Bus 09
	0.92305	0.95276

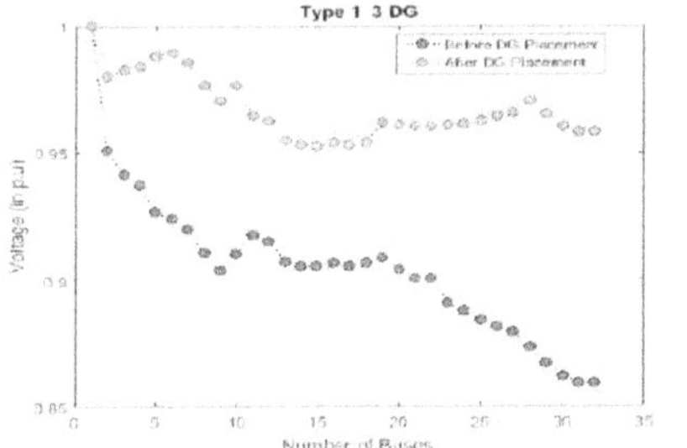

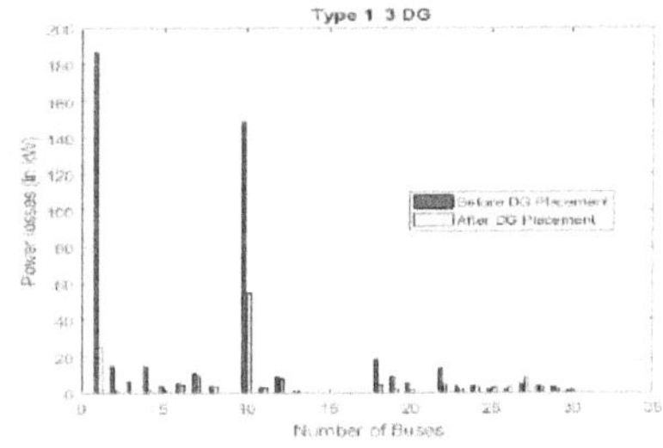

Figure 4.88: Voltage profile and Power losses of the system for penetration of Hybrid DG-scenario 5

Case-3: Optimal placement of Solar and Wind (Hybrid) DG units

The results obtained from the developed MSFA-PSO algorithm for case-3 is tabulated in table.4.5. The system losses reduced to 90.61 kW from 477.99 kW and percentage losses are reduced to 81.04 %. Minimum voltage before DG placement is 0.85952 p.u at bus 32 and after DG placement the minimum voltage obtained to be 0.96921 p.u.at bus 32.

Table 4.75: Optimal allocation of Solar and wind hybrid DG by MSFA-PSO

Particulars	MSFA-PSO Hybrid DG
1: OLDG	Bus 08 Bus 21
2: ODGS	1491kW, 3505.3kW, -1078.8 kVAR
3: BCPL (kW)	477.09
4: PLDG (kW)	90.61
5: PLR (%)	81.04
6: Min V w/o DG (p.u.)	Bus 32 0.85952
7: Min V W DG (p.u.)	Bus 32 0.96921

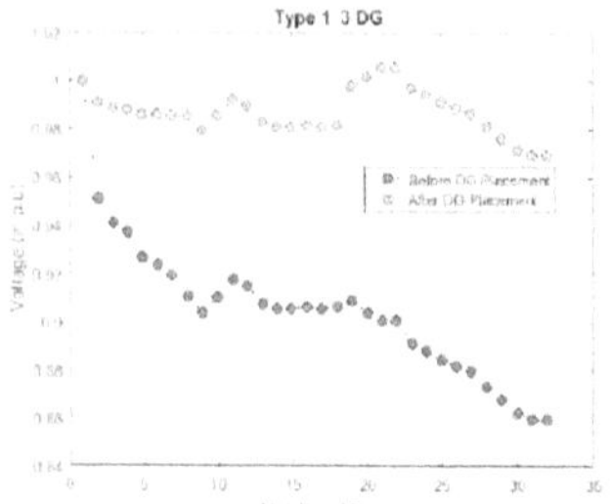
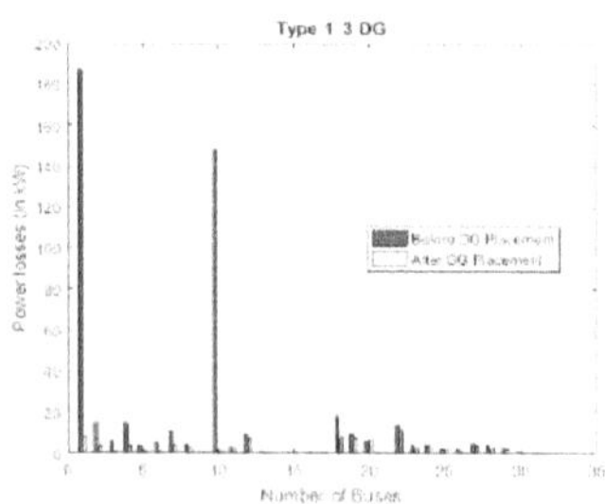

Figure 4.89: Voltage profile and Power losses of the system for penetration of Hybrid DG

Table 4.76: WSOF values at different buses after integration of solar and wind hybrid DG at different penetration levels

Bus number	WSOF at 10%	WSOF at 20%	WSOF at 30%	WSOF at 40%	WSOF at 50%
1	0	0	0	0	0
2	0.170734	0.232612	0.311729	0.378111	0.424516
3	0.032175	0.04732	0.067436	0.088057	0.480406
4	0.032394	0.04763	0.067902	0.08867	0.490475
5	0.032976	0.048464	0.069096	0.090209	0.508075
6	0.033149	0.048757	0.069484	0.090712	0.500335
7	0.033387	0.049113	0.069982	0.091365	0.232425
8	0.033921	0.049868	0.071069	0.0928	0.235954
9	0.034409	0.050586	0.072051	0.09405	0.23896
10	0.033883	0.049833	0.071037	0.092716	0.235788
11	0.256931	0.341904	0.443414	0.516417	0.383221
12	0.073735	0.347934	0.519723	0.521014	0.169448
13	0.074693	0.369126	0.541794	0.523541	0.171621
14	0.074996	0.123011	0.185558	0.240505	0.172255
15	0.075001	0.123021	0.185573	0.240518	0.172267
16	0.074828	0.12273	0.444044	0.239977	0.171897
17	0.074992	0.122998	0.193582	0.24048	0.172241
18	0.074792	0.122724	0.193184	0.239955	0.171858
19	0.354757	0.357241	0.382882	0.454462	0.450288
20	0.38594	0.390184	0.408221	0.474595	0.459658
21	0.408064	0.415304	0.425618	0.484933	0.457501
22	0.107778	0.132614	0.178556	0.230831	0.21708
23	0.442918	0.454087	0.453931	0.50434	0.459985

24	0.385823	0.499589	0.465917	0.472965	0.362945
25	0.400912	0.519482	0.469192	0.45399	0.310397
26	0.41707	0.539842	0.448871	0.373625	0.30652
27	0.424125	0.548527	0.435409	0.326994	0.375696
28	0.437392	0.565481	0.42532	0.314915	0.45669
29	0.18078	0.532021	0.236558	0.318451	0.356039
30	0.182508	0.45287	0.302593	0.321417	0.359278
31	0.183409	0.280422	0.304033	0.322932	0.360961
32	0.183428	0.280442	0.304042	0.322948	0.360988

4.5 Summary

A hybrid approach combining the advantages of both SFA and PSO method is proposed for the optimal allocation of different DG's and their combinations. PSO is used for optimal sizing and SFA is used for the optimal location. The analysis is carried out for optimal location of DG for the corresponding penetration level under normal operation and optimal allocation under heavy load condition of the test system. The analysis is carried out on the IEEE 33bus, IEEE 69bus test system and 32 bus practical test system for different scenarios. Also, the comparison is done for the different type of DG's and their allocation with the PSO method shown in the comparison table. It is evident that the percentage losses in the system are reduced and minimum voltage at the bus is increased under normal condition. During heavy load, the DG will compensate the entire additional load to give a better efficiency by nullifying voltage sag at the affected nodes.

CHAPTER 5

5 Conclusions

The work in this research report is based on proposing new algorithms for different penetration of renewable DGs for different conditions and optimal allocation in electrical power system with different objectives which address the practical problems of utility. As the DGs is a real power generating source, sometimes reactive power consuming and generating sources connected in the system, it is possible to analyze the impacts caused by the integration of DG's in the system. Here, all the nodes of the system are not suitable for connection of the distributed generator. The analysis of the system is done on the basis of the penetration and method of integration of the chosen DG, the problem of siting of the generators is a very important part for enhancing the system operation when renewable DGs are connected. The major observations of this thesis are,

1. An analytical method is developed for large penetration of renewable DGs and optimal allocation of the system for the improvement of the voltage profile and power loss minimization.

2. Development of an intelligent algorithm for different penetration of renewable DGs with optimal allocation in the network, for the improvement of the voltage profile and reduction in the power loss under normal and heavy loaded conditions with power quality indices using the Modified shuffled frog leap algorithm.

3. Development of a Meta-heuristic hybrid method for large penetration of renewable DGs considering different conditions for the penetration with optimal allocation using shuffled frog leap and particle swarm optimization algorithm. For the minimization of the losses and improvement in the voltage profile under heavy different scenarios with hybrid power system.

The salient conclusions of respective techniques are given below,

5.1 Salient conclusions

5.1.1 Analytical approach of optimal allocation of DG unit in the electrical network

The voltage deviation problem is a challenge faced in many utilities, an algorithm for the optimal placement according to certain penetration level for solar and wind powered DGs is developed. The main objective of this being simultaneous reduction of the power loss and weighted sum of voltage index (WSVI) by reducing logarithmic voltage deviation index

(LVDI) and voltage profile index in the system considered. A new index named "WSVI" is defined and used for the selection of the optimal location of the DG. The algorithm is executed in multiple steps where the LVDI and Voltage profile index (VPI) are determined at the first step then WSVI is calculated. The bus with the maximum value of the index is selected. In the next stage the penetration of the DG is decided and the DG's are placed accordingly. Later the analysis is evaluated by find outing the value of Voltage profile Improvement Index (VPII), This method is applied on the standard IEEE 33 bus test system and practical 32 bus test system.

Upon analysis of the results of various test systems, it is noted that the integration of solar distributed generators increases the capacity of delivering power compared to the wind power generation. Comparison is done based on better voltage profile and reduced power losses.

5.1.2 Development of an intelligent algorithm for large penetration and allocation of renewable DGs in the network using Modified frog leap algorithm

In this chapter, the 'problem of allocation of DG's" is formed as a multi-objective function and the solution is obtained using and intelligent method with modification called as modified shuffled frog leap algorithm. This algorithm includes the objective of real power loss reduction and weighted sum of voltage profile index for improving the voltage profile. The algorithm is coded in the MATLAB platform and applied on the standard IEEE 33 bus test system and practical 32 bus system. Upon analysis of the results of various test systems, it is noted that the integration of solar and wind powered DGs leads to improve the voltage profiles and reduce the losses under normal and heavy loaded condition. This method with hybrid power generation yields better results upon comparison with other methods, it is observed that the technique simultaneously reduces the real power losses and also improves the voltage profile of the system by reducing the deviation index. It is seen that the losses are reduced more than 60% while the voltage profile is improved to the range above 0.95 whereas the base case voltage gives the value of 0.92 p.u.

5.1.3 Development of a meta-heuristic algorithm for large penetration of renewable DGs under different conditions with optimal allocation using Modified flower pollination algorithm

In this chapter an algorithm for determining the optimal locations and penetration based on the different cases and scenarios are considered. A hybrid approach combining the advantages of both SFA and PSO method is proposed for the optimal allocation of different DG's and their combinations. PSO is used for optimal sizing and SFA is used for the optimal location. The analysis is carried out for optimal location of DG for the corresponding penetration level under normal operation and optimal allocation under heavy load condition of the test system. Analysis is made for the system so that from it the real power losses can be reduced and for the network the voltage profile can be improved at the same time. In MATLAB the coded algorithm along with other used algorithms are tested on the practical 32 bus system and standard IEEE bus test system for distributed generation of solar, wind and hybrid energies. From this algorithm the outcomes obtained are contrasted with the outcomes which are extracted from literature and other methods. It is also shown that by the suggested method there is a maximum power loss. It can be used as an examination tool for renewable DG allocations.

5.2 Future scope

The work in this dissertation opens the path for future investigation:

a) The load in the network is assumed to be static and considered the change only based on the fixed interval of time during the optimization process. But in the practical system, the loads are dynamic in nature. Hence the work can be carried forward for dynamic loading.

b) The load growth of the network is not considered. Work can be extended for load growth over a period of time and can extended for the analysis of load forecasting.

c) The major concentration in the research was the power loss, voltage profile and power quality indices based on the heavy load and losses under fault conditions. This can be extended considering other factors like harmonic analysis.

d) The algorithms in the dissertation shall be extended for the micro grid planning considering different DG's with other constraints.